SMITHSONIAN
TREASURES OF
THE **NATIONAL
AIR AND SPACE
MUSEUM**

AMERICAN
Ford
U.S. MAIL
EASTERN AIR LINES
EASTERN
FORD
ENTERED SERVICE
1928
5-AT TRI-MOT
DOUGLAS
DC-3
NORTHROP
4A ALPHA
BOEING
247-D
AMERICA BY AIR
ENTRANCE
AMERICAN

SMITHSONIAN TREASURES OF THE **NATIONAL AIR AND SPACE MUSEUM**

Tony Reichhardt
with the National Air
and Space Museum

Smithsonian Books
WASHINGTON, DC

© 2023 by Smithsonian Institution
All rights reserved. No part of this publication may be reproduced or transmitted in any form or by any means, electronic or mechanical, including photocopying, recording, or information storage or retrieval system, without permission in writing from the publishers.

Smithsonian Books
Director: Carolyn Gleason
Senior Editor: Jaime Schwender
Assistant Editor: Julie Huggins

Edited by Sharon Silva
Designed by Gary Tooth / Empire Design Studio

National Air and Space Museum
John and Adrienne Mars Director: Christopher U. Brown
Publications Team: Jeremy Kinney, Polly McKenna-Cress, Michael J. Neufeld, and F. Robert van der Linden Image Research: Melissa A. N. Keiser, Eric Long, Jim Preston, and Carolyn Russo

This book may be purchased for educational, business, or sales promotional use. For information, please write: Special Markets Department, Smithsonian Books, P.O. Box 37012, MRC 513, Washington, DC 20013

Library of Congress Cataloging-in-Publication Data is available upon request.

Paperback ISBN: 978-1-58834-735-0

Printed in China, not at government expense
27 26 25 24 23 1 2 3 4 5

For permission to reproduce illustrations appearing in this book, please correspond directly with the owners of the works, as seen on page 118. Smithsonian Books does not retain reproduction rights for these images individually or maintain a file of addresses for sources.

I: *Beech Staggerwing C17L (page 32) and the Space Shuttle* Discovery *(page 96);* ***II:*** *view of the* America by Air *exhibit at the National Air and Space Museum;* ***III:*** *Bell X-1 (pages 52–53) and Charles Lindbergh's* Spirit of St. Louis *(pages 26–27);* ***V:*** *Mars Exploration Rover (page 103).*

Table of Contents

COLLECTING THE PAST, **INSPIRING THE FUTURE**

Air and space. One is familiar, accessible, the realm of wind and clouds. The other is distant and alien, in some ways more imaginary than real. Yet huge strides in understanding both domains have happened only recently, and at about the same time. Humans, long confined to land and sea, learned in the twentieth century to travel routinely above and beyond Earth's surface in airplanes and spacecraft.

It should come as no surprise, given its mission to advance and disseminate knowledge, that the Smithsonian embraced the possibilities of flight almost from its founding in 1846. The institution's first secretary, Joseph Henry, recognized the value of early balloon-borne research and,

"Rocket Row," on display outside the Arts and Industries Building, ca. 1961. From left to right: the Jupiter C, which launched Explorer I, the first US satellite; the Vanguard; the Polaris A-1 submarine-launched missile; and the Atlas, which launched Mercury orbital missions.

Before the current National Air and Space Museum building was finished in 1976, treasures like the 1903 Wright Flyer (hanging overhead) and Apollo 11 Command Module (front right) were displayed inside the Smithsonian's Arts and Industries Building.

in 1859, provided the meteorological equipment carried aboard *Smithsonian*, a balloon launched by aeronaut John Wise. Two years later, Henry introduced President Abraham Lincoln to pioneering balloonist Thaddeus Lowe, who made several tethered ascents on the National Mall near the current site of the National Air and Space Museum to demonstrate how balloons could be used to spy on enemy troops during the Civil War.

The Smithsonian's third secretary, Samuel Langley, was even more enamored of flight and made important contributions himself to the science of aeronautics. His Aerodrome Number 5—the first unpiloted, engine-driven, heavier-than-air craft of substantial size—is in the museum today. Although Chinese kites exhibited at the 1876 Philadelphia Centennial International Exhibition are considered the oldest human-made flying objects in the Smithsonian's collection, it was Langley who began systematically acquiring the institution's first aviation-related artifacts,

The museum's Steven F. Udvar-Hazy Center, adjacent to Dulles International Airport in the Virginia suburbs of Washington, DC, opened to the public in 2003.

starting with a one-horsepower steam engine for aircraft built by British inventor John Stringfellow, which Langley bought for twenty-five pounds in 1889.

More acquisitions and donations followed, thanks in part to sharp-eyed museum staff who kept on the lookout for significant new technical developments. In 1927, aviation curator Paul Garber persuaded the Smithsonian secretary to send a telegram to Charles Lindbergh asking him to donate his *Spirit of St. Louis* following his solo transatlantic flight. Eight years later, the museum acquired a liquid-propellant rocket built by American scientist Robert Goddard, whose pioneering research the Smithsonian had supported. Lindbergh, an advocate for Goddard, had urged him to donate the vehicle to the institution.

As the aerospace collection grew, the need for a dedicated building became clear. In 1919, the Smithsonian rebuilt a wartime tin shed behind the Castle to show its aeronautical collections. Important artifacts, such as the *Spirit*, were exhibited in the neighboring Arts and Industries

Alejandro Otero's stainless steel sculpture, Delta Solar, *shines at the west end of the National Air and Space Museum on the Washington, DC, National Mall.*

Building. The National Air Museum became a separate bureau of the Smithsonian in 1946. An outdoor Rocket Row followed in the late 1950s, and in 1966, the name was changed to National Air and Space Museum to acknowledge the growing number of objects related to space exploration.

In 1976, in time for the nation's bicentennial and just a few years after the Apollo moon program ended, a new building dedicated to air and space opened on the National Mall that could display very large objects indoors, from a Douglas DC-3 to the Skylab orbital workshop. In 2003, for the one hundredth anniversary of the Wright brothers' first flight, a second campus opened in the Virginia suburbs of Washington, DC—the Steven F. Udvar-Hazy Center—which also houses the museum's restoration facility. An even greater number of large aircraft and spacecraft, from the SR-71 Blackbird to Space Shuttle *Discovery*, are on display here.

Today, the National Air and Space Museum holds in trust more than sixty thousand artifacts and millions of additional objects, from letters and manuscripts to technical drawings, photographs, and works of art. Together, they document the field of aerospace in all its richness and variety. Many are on long-term loan from individuals, companies, and government agencies. In turn, the Smithsonian loans objects in its collection to other museums and also displays them in its public buildings, traveling exhibitions, and online galleries.

Beginning in 2018, the downtown Washington museum has undergone a monumental multiyear renovation that will reimagine all twenty-three exhibitions. Like the field of aerospace itself, the National Air and Space Museum keeps evolving.

TAKING FLIGHT

Flying has always figured in our myths and dreams, from angels to Icarus. Even before people took to the sky themselves, they flew kites and launched toy helicopters. By the end of the eighteenth century, aeronauts were floating aloft in balloons, followed by the first winged gliders. Then, at the dawn of the twentieth century, inventors hit on the right combination of technologies to achieve controlled, powered, heavier-than-air flight—the first practical airplanes. The National Air and Space Museum collection includes artifacts from the long prehistory of flight as well as some of the most iconic aircraft from aviation's first decade.

(Artwork: *Balloon over City*, artist unknown, 1783)

Kites were the first human-made objects to achieve sustained flight. This **Chinese dragon kite** was originally shown at the Centennial International Exposition in Philadelphia in 1876 and was one of the first aeronautical objects in the museum's collection.

Following the first launches of **hot-air balloons** over Paris in the autumn of 1783, crowds across Europe, and soon the world, were captivated by the new age of flight. Balloon motifs appeared on everything from jewelry to furniture to this painted wooden case for ladies' dance cards, or *carnets de bal*.

This quarter-plate glass ambrotype made by an unknown photographer shows preparations for a **balloon ascension** by John Steiner at Erie, Pennsylvania, on June 18, 1857. The scene, picturing a ring of people surrounding the balloon as it's being inflated, is thought to be the earliest dated photograph of a balloon flight in the United States.

Sir George Cayley built the world's first **hand-launched model glider** in 1804, with separate systems for lift and control. A quarter-scale reproduction of the five-foot original is in the museum.

Inflatable, engine-driven airships preceded the invention of the airplane at the start of the twentieth century. Alberto Santos-Dumont (shown in the gondola of one of his airships), the son of a wealthy Brazilian coffee farmer, made headlines piloting his small, hydrogen-filled craft around Paris. This lightweight **Clement V-2 engine** powered Airship No. 9 in 1903. Santos-Dumont also made Europe's first public airplane flight with his 14-bis in 1906.

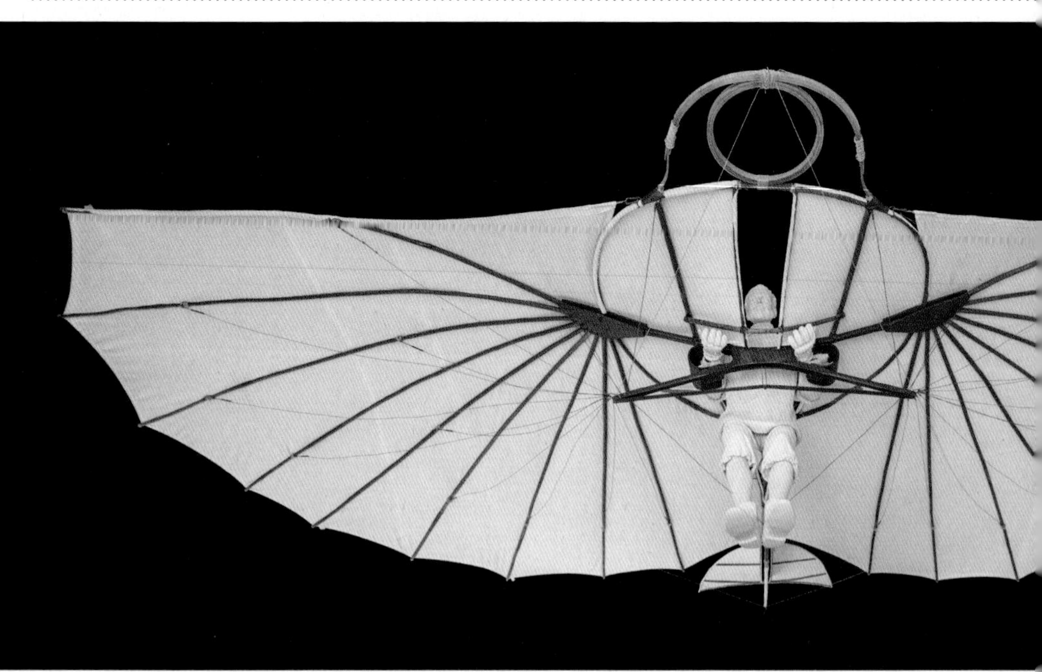

"Of all the men who attacked the flying problem in the nineteenth century, Otto Lilienthal was easily the most important," claimed Wilbur Wright. Between 1891 and 1896, the German aviation pioneer designed and built a series of full-size gliders, making nearly two thousand flights himself. The museum's 1894 model is one of only six **Lilienthal gliders** still in existence.

In 1891, astronomer and Secretary of the Smithsonian Samuel P. Langley began experiments with a series of large, tandem-winged models he called aerodromes, which were equipped with small steam and gasoline engines. On May 6, 1896, near Quantico, Virginia, **Langley's Aerodrome Number 5**, now in the museum, made the world's first successful flight (pictured) of an unpiloted, engine-driven, heavier-than-air craft of substantial size. A full-scale piloted version failed in 1903, just weeks before the Wrights' first successful airplane flight.

The 1903 **Wright Flyer**, with Orville Wright at the controls, takes off from the foot of the big Kill Devil Hill near the village of Kitty Hawk, North Carolina, at 10:35 on the morning of December 17. Orville's historic first flight lasted just twelve seconds and covered a distance of 120 feet. The longest of the four flights completed that day covered 852 feet in fifty-nine seconds.

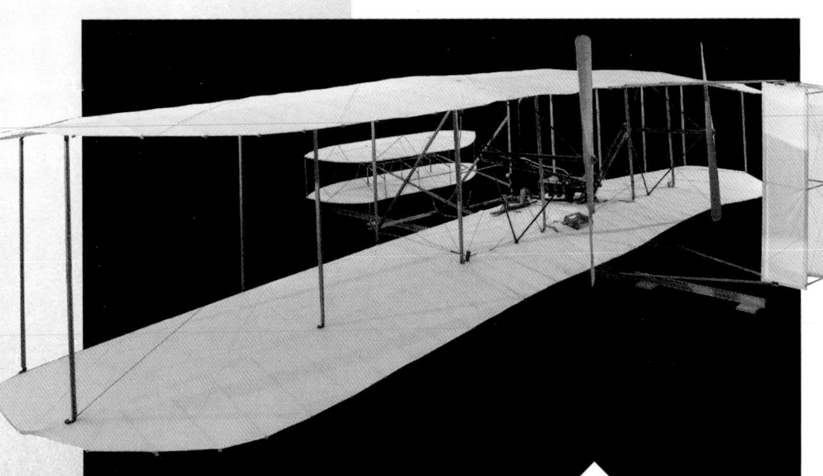

Self-taught engineering geniuses from Dayton, Ohio, Wilbur and Orville Wright began their quest to fly, starting with kites, in 1899. Over the next several years, the brothers built and tested a series of three gliders and three flying machines equipped with engines. This vehicle achieved the first controlled, sustained, powered human flight on December 17, 1903. The airplane never flew again; it was donated to the Smithsonian forty-five years later.

The Wright brothers used this **stopwatch** to time their first airplane flights at Kitty Hawk.

In the decade following the Wrights' first flights, other pioneers on both sides of the Atlantic produced their own designs and set new records. In 1909, engineer and aviator Louis Blériot of France crossed the English Channel by air. Glenn Curtiss of Hammondsport, New York—a bicycle maker like the Wright brothers as well as a motorcycle racer—came up with his **Headless Pusher** design, so named because the propellers push rather than pull the airplane, and there was no forward elevator as in the Wrights' airplanes. Curtiss briefly became the best-known aviator in America after his mile-long public flight in July 1908.

In 1910, the year this photo was taken, **Curtiss** flew from Albany to New York City, the first long-distance flight between major cities. His company also built and flew some of the first practical seaplanes.

The Curtiss Model D was the standard airplane flown by pilots on the Curtiss exhibition team, including **Blanche Stuart Scott**, the only woman to receive instruction from Glenn Curtiss himself. Scott was the first American woman to fly an airplane solo, with a short hop on September 6, 1910.

Curtiss's experimental **June Bug**, with its fixed-pitch wood propeller, won the first aeronautical prize awarded in the United States, the Scientific American Cup, in 1908. It flew 5,090 feet in a straight line, beating the required distance of one kilometer.

The French-built **Gnome rotary engine**, introduced in 1909, was popular among early airplane builders.

Early aviators were always looking for daring new ways to thrill an air-show crowd. **Georgia Ann "Tiny" Broadwick** was just fifteen years old when she jumped from a hot-air balloon at the 1908 North Carolina State Fair. When famed stunt flyer and airplane manufacturer Glenn L. Martin proposed that she parachute from an airplane, she became the first woman to do so on June 21, 1913. Broadwick donated her parachute to the museum in 1964.

The 1909 **Wright Military Flyer** is the world's first military airplane. In 1908, the US Army Signal Corps sought competitive bids for a two-seat observation aircraft. Flight trials with the Wrights' entry began at Fort Myer, Virginia, that September. After several days of successful flights, tragedy struck when the Wright Military Flyer crashed, injuring Orville Wright and killing his passenger, US Army observer First Lieutenant Thomas E. Selfridge, who became the first fatality in a powered airplane. The Wrights returned to Fort Myer the following June with a new aircraft that met all of the army's requirements during trials. It was given to the Smithsonian in 1911.

SIGNAL CORPS SPECIFICATION, NO. 486.

ADVERTISEMENT AND SPECIFICATION FOR A HEAVIER-THAN-AIR FLYING MACHINE.

TO THE PUBLIC:

Sealed proposals, in duplicate, will be received at this office until 12 o'clock noon on February 1, 1908, on behalf of the Board of Ordnance and Fortification for furnishing the Signal Corps with a heavier-than-air flying machine. All proposals received will be turned over to the Board of Ordnance and Fortification at its first meeting after February 1 for its official action.

Persons wishing to submit proposals under this specification can obtain the necessary forms and envelopes by application to the Chief Signal Officer, United States Army, War Department, Washington, D. C. The United States reserves the right to reject any and all proposals.

Unless the bidders are also the manufacturers of the flying machine they must state the name and place of the maker.

Preliminary.—This specification covers the construction of a flying machine supported entirely by the dynamic reaction of the atmosphere and having no gas bag.

Acceptance.—The flying machine will be accepted only after a successful trial flight, during which it will comply with all requirements of this specification. No payments on account will be made until after the trial flight and acceptance.

Inspection.—The Government reserves the right to inspect any and all processes of manufacture.

GENERAL REQUIREMENTS.

The general dimensions of the flying machine will be determined by the manufacturer, subject to the following conditions:

1. Bidders must submit with their proposals the following:
   (*a*) Drawings to scale showing the general dimensions and shape of the flying machine which they propose to build under this specification.
   (*b*) Statement of the speed for which it is designed.
   (*c*) Statement of the total surface area of the supporting planes.
   (*d*) Statement of the total weight.
   (*e*) Description of the engine which will be used for motive power.
   (*f*) The material of which the frame, planes, and propellers will be constructed. Plans received will not be shown to other bidders.

The **US Army's specifications** for its military airplane included the requirement that it be able to fly at forty miles per hour in still air.

1909
AERONAUTICAL
TRIALS
FORT MYER
VA.

Spectators needed a **pass** to get into the Fort Myer flight trials.

AVIATION GOES TO WAR

Airplanes had been used in warfare before 1914, but World War I proved their lasting value. Aerial photography allowed commanders to survey the battlefield in near real time. While bombers attacked from above, fighters powered by ever-more-capable engines jousted in the skies. Pilots who entered the war never having seen an airplane invented dogfighting tactics still used today. This painting in the museum's collection by French-born artist Henri Farré, who accompanied pilots on combat missions, depicts an aerial battle won by pilot Georges Guynemer, one of the highest-scoring French aces, as the new knights of the air were called.

(Artwork: *Ailes Glorieuses, The 45th Victory of Guynemer*, Henri Farré, 1917)

The **first American pilots** to fight in World War I flew for France in a squadron known as the Lafayette Escadrille. The thirty-eight American members of this all-volunteer unit joined before the United States entered the war in 1917 and served under French officers. Many, like Norman Prince and William Thaw, came from wealthy backgrounds. Their experience in the early years of the war became invaluable to the US Army Air Service formed in 1918.

Members of the **Lafayette Escadrille** pose with Whiskey and Soda, the unit's lion cub mascots, in France in 1917.

Henri Farré painted this portrait of **Raoul Lufbery,** an American citizen born in France who became the Lafayette Escadrille's first ace following his fifth aerial kill in 1916. After the United States joined the war, Lufbery was transferred to the US Army as a pilot instructor and unit commander. He was killed in combat in 1918, after seventeen previous victories.

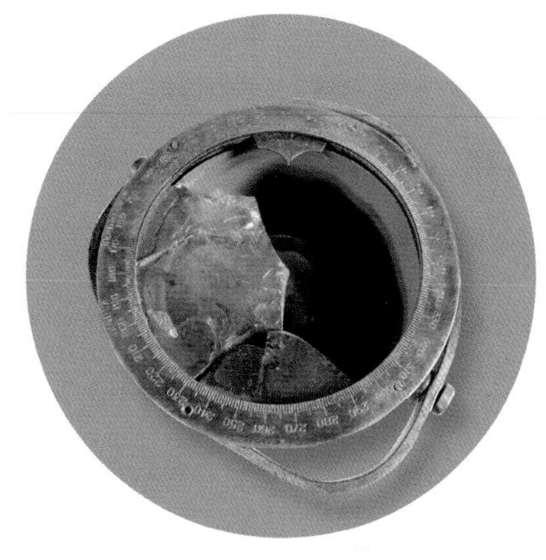

This **aircraft compass** was recovered from the Nieuport 28 in which Lufbery was mortally injured on May 19, 1918, over Maron, France.

Kiffin Rockwell was the first American pilot to shoot down an enemy aircraft in World War I. Awarded France's **Médaille Militaire** for bravery, he was killed in action in 1916.

The French-built **SPAD XIII** was among the most successful fighter aircraft of World War I. It was widely used by French and American forces as well as by Italy, Belgium, Russia, and Great Britain. The single-seat, single-engine biplane was noted for its rugged construction and ability to dive at high speeds, which made it especially good in dogfights. The museum's SPAD was flown by American pilot Arthur Raymond "Ray" Brooks.

MIT graduate **Ray Brooks** flew 120 missions in four different aircraft for the US Army's Twenty-Second Aero Squadron. He named each of his planes *Smith* in honor of his fiancée, who attended Smith College. The museum has Smith IV. The last surviving American World War I ace, Brooks died in 1991.

In his **wartime diary**, now in the Smithsonian, rookie pilot Brooks recalled asking his flight commander about his first encounter with antiaircraft fire. "What did you do when your first antiaircraft burst went off?" I asked. "Jumped about a foot out of my seat," the commander answered.

Paul R. Stockton enlisted in the US Army Signal Corps in 1906 and served as commanding officer of the Twelfth Aero Squadron of the American Expeditionary Forces during World War I. His **scrapbook**, now in the museum's collection, details his and his comrades' service before, during, and immediately after the war.

Introduced in 1917, the French-built **Nieuport 28C.1** was the first fighter aircraft to serve with an American fighter unit under American command and in support of US troops. The museum's Nieuport bears the Ninety-Fourth Aero Squadron's Hat in the Ring insignia, indicating that Uncle Sam had joined the war.

When Germany's **Fokker D.VII** appeared on the Western Front in April 1918, Allied pilots at first underestimated it. But the ability of the single-seat fighter aircraft to "hang on its propeller" while firing from underneath opponents made it a feared enemy. The Fokker D.VII in the museum collection was built by Albatros.

Captain Edward "Eddie" Rickenbacker, commander of the Ninety-Fourth Aero Squadron, was the United States' top scoring ace of World War I, with twenty-six confirmed victories. He was later awarded the Medal of Honor for single-handedly attacking seven enemy airplanes and shooting down two of them on September 25, 1918.

Rickenbacker wore this wool **uniform coat**, with pilot's badge on the left breast, while serving in France during World War I.

Aerial reconnaissance was a vital strategic tool during World War I. The two-seat **De Havilland DH-4** designed in Britain was used as both a bomber and a reconnaissance plane, with one crew member taking pictures from the open cockpit.

US Army crews flying some of the war's most dangerous missions took millions of **overhead shots** like this one, showing irregularly shaped trenches on the battlefield.

Kodak's A-1 (pictured) and A-2 handheld cameras, both of which are in the museum's collection, were among the most widely used models for aerial reconnaissance during World War I.

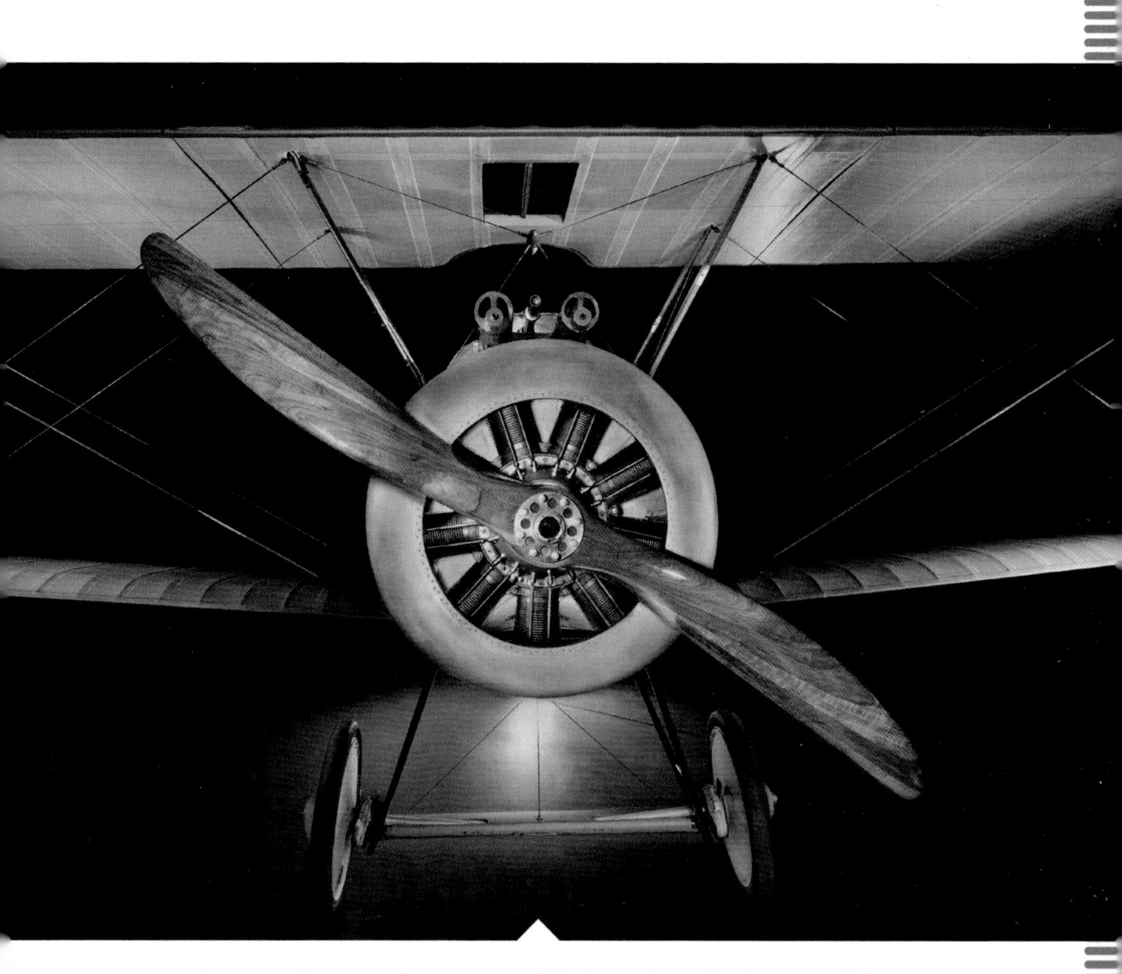

Among the most famous World War I aircraft, the **Sopwith Camel** got its name from the hump-like cowling over the two Vickers machine guns ahead of the cockpit. Camels downed more enemy aircraft than any other Allied fighter, but their instability made them dangerous for novice pilots, and almost as many were killed in accidents as died in combat. The museum's Camel served with the No. 10 Squadron of the Royal Naval Air Service.

THE SHRINKING **GLOBE**

By the time World War I ended in 1918, more than two hundred thousand airplanes had been produced and a whole generation of pilots had been trained. The surplus flying machines found all sorts of new applications, from Hollywood stunt flying to delivering air mail to exploring remote regions of the globe. Racing competitions and traveling barnstorming exhibitions pushed technology and piloting skill to the edge in the 1920s and 1930s and introduced millions of spectators to the wonders of flight.

(Artwork: *Spirit of St. Louis*, Robert C. Korta, ca. 1960)

In May 1919, eager to show that airplanes were capable of flying long distances, the US Navy sent three Curtiss flying boats, with crews drawn from the navy and coast guard, on a voyage across the Atlantic. The twenty-four-day journey, which began in Rockaway Beach, New York, and ended in Portugal, was broken into several legs. Only one of the air-planes, the **NC-4**—shown arriving in Lisbon harbor—flew the whole way, but the feat was celebrated around the world.

The crew of the NC-4 used a **radio receiver** like this one on its transatlantic voyage.

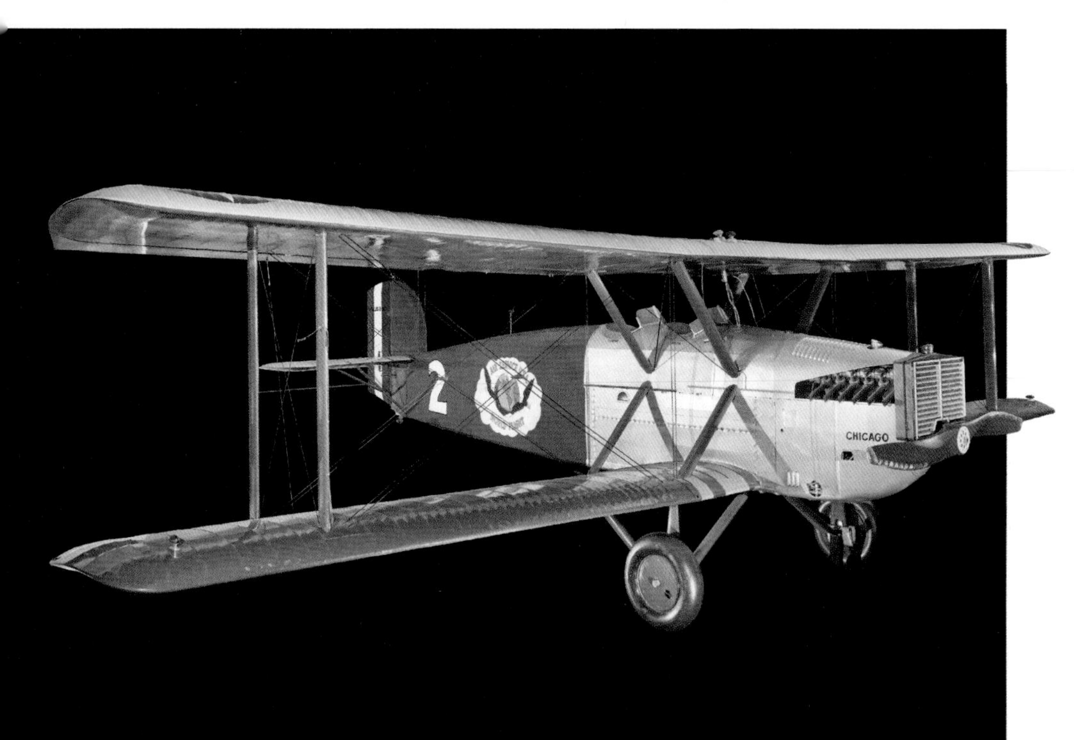

Following successful transatlantic flights by Americans and Europeans, the next big challenge was circling the globe. In 1924, the US Army Air Service sent four World Cruisers—the *Seattle*, the *Boston*, **the *Chicago***, and the *New Orleans*, all specially built by the Douglas Aircraft Company—on the arduous 27,553-mile journey around the world. Only the last two completed the flight, which took 175 days and a total flying time of 371 hours, 11 minutes, with seventy-four stops along the way. The *Chicago* is now in the Smithsonian.

Crew members on the World Cruisers' voyages carried stuffed monkey dolls as mascots. **Maggie**, signed by pilot Lieutenant Leigh Wade, rode on the *Boston*.

In the spring of 1927, Charles Lindbergh, a former barnstormer and air mail pilot from Minnesota, competed for the $25,000 Orteig Prize for the first nonstop flight between New York and Paris. With financial backing from several Saint Louis businessmen, he placed an order with Ryan Airlines in San Diego for a single-seat, single-engine, high-wing monoplane that he designed with Ryan's Donald Hall. Taking off from New York on May 20, the ***Spirit of St. Louis*** landed at Le Bourget field thirty-three hours later. Lindbergh became an instant celebrity. A year later, he delivered the airplane to the Smithsonian—its final flight.

Lindbergh and his wife and fellow pilot, Anne Morrow Lindbergh, became **ambassadors for aviation** in the 1930s. In a modified Lockheed Sirius, they embarked on two long ocean voyages—first to the Far East, then across the North and South Atlantic— to scout routes for commercial airliners.

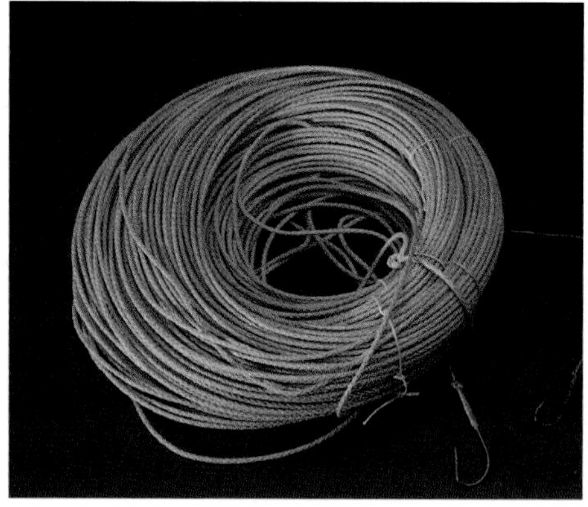

In case his plane went down over the ocean, Lindbergh carried a basic survival kit, including this **fishing line and hook**.

Anne Lindbergh's **embroidered sealskin boots** were useful for stops in cold places. Her pith helmet, also in the Smithsonian, was standard wear in the tropics.

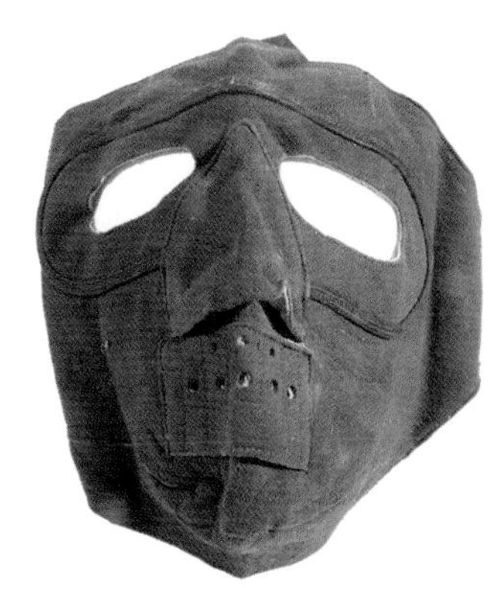

Charles Lindbergh wore this **face mask** to keep warm during the couple's five-month, four-continent circumnavigation of the Atlantic in 1933.

The **Curtiss JN-4D Jenny** is almost synonymous with American aviation in the 1920s. After serving as wartime trainers, the relatively cheap biplanes were widely used by barnstormers and air mail pilots in the postwar years. The Smithsonian acquired this Jenny in 1918 directly from the US War Department. It is one of the finest remaining examples of this classic airplane.

Because she was unable to get a pilot's license in Jim Crow America, Chicago manicurist **Bessie Coleman** (pictured here with a Curtiss Jenny) traveled to France in 1921 to earn the first license awarded to an African American woman. Back in Chicago, she performed at air shows as a barnstormer, flying daring loops and rolls—but only for crowds that, by her insistence, were desegregated. At the time of her death in 1926 (she accidentally fell from an airplane flown by another pilot), Coleman was planning to open a flight school for African American pilots.

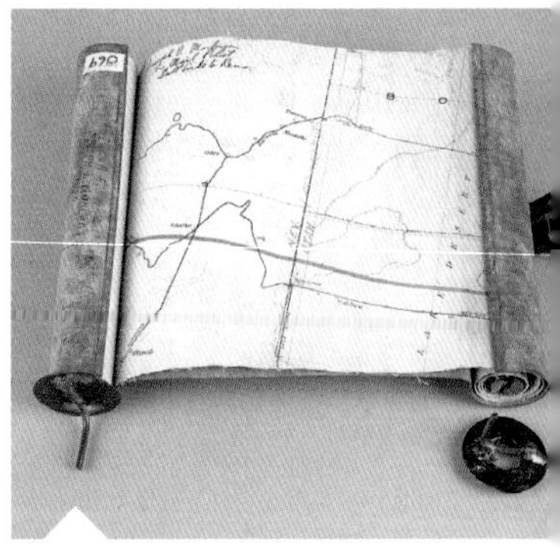

Air mail pilots used "**knee board" maps** strapped to their legs to navigate along their route, turning knobs to scroll the map as they flew. This one helped pilot Joseph L. Mortensen find his way from Salt Lake City, Utah, to Reno, Nevada, in 1920.

In 1925, the army selected a young test pilot, Lieutenant James Doolittle, to fly this **Curtiss R3C-2** in a race to win the Schneider Trophy. Doolittle, who won renown seventeen years later for his wartime raid on Tokyo, piloted the R3C-2 to victory with an average speed of 232.6 miles per hour over Maryland's Chesapeake Bay. The next day, he set a new sea-plane world record speed of 245.7 miles per hour over a straight course. This plane is now a part of the museum's collection.

Robert Goddard, a Massachusetts college professor who had dreamed of spaceflight since he was a teenager, conducted pioneering experiments with liquid-fueled (gasoline and liquid oxygen) **rockets** in the 1920s and 1930s. Early research funding came from the Smithsonian, which published his seminal work, *A Method of Reaching Extreme Altitudes*, in 1920. Goddard's wife, Esther, took this picture of her husband on March 8, 1926, at a farm in Auburn, Massachusetts. Eight days later, he successfully launched the world's first liquid-fueled rocket to an altitude of forty-one feet.

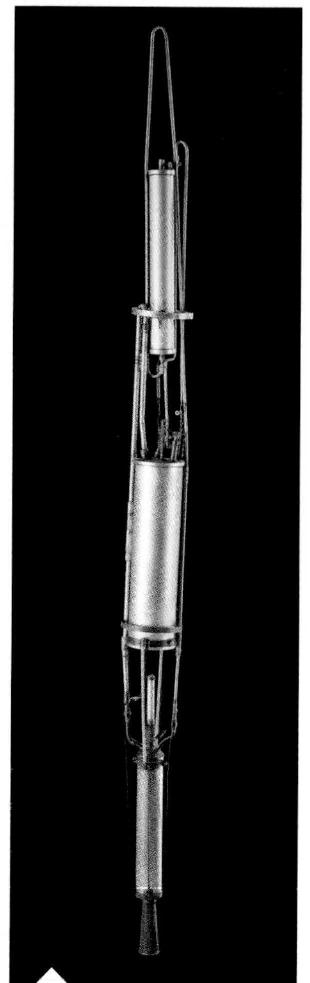

Built in May 1926, this is the oldest-surviving **liquid-propellant rocket** in the world. For this version, Goddard placed the motor at the base instead of in the nose, as in his successful rocket of two months earlier. Although this rocket wasn't able to launch, it likely includes the nozzle from the historic March 1926 vehicle.

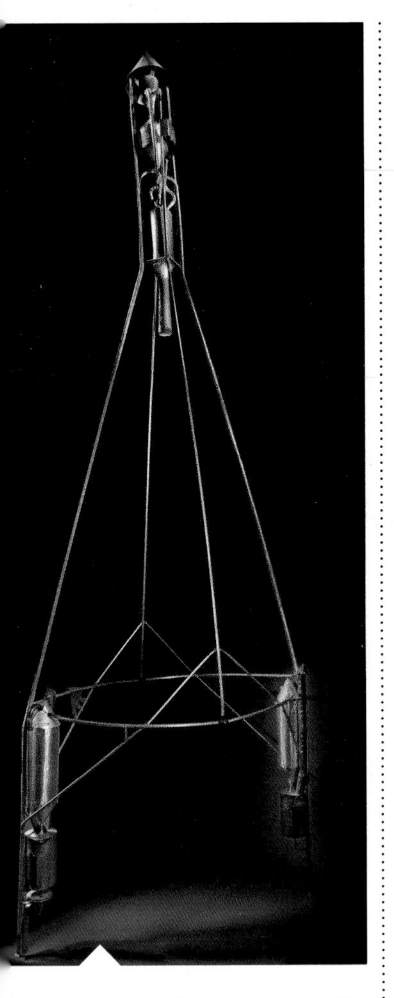

On December 26, 1928, on his fourth try, Goddard launched his **Hoopskirt rocke**t, named for the ladies' dress design, from a site near Worcester, Massachusetts. Rather than aim for a streamlined shape, he wanted to test his rocket motor in as light a vehicle as possible. The flimsy structure was smashed by the flight. The Smithsonian's Hoopskirt rocket is a reconstruction.

The is the nose cone of Goddard's 1935 **A-series rocket**, likely the same one he tried to launch on September 23, 1935, at Roswell, New Mexico, in an attempt to demonstrate its capabilities to supporters Charles Lindbergh and Harry Guggenheim, whose family's foundation funded Goddard's experiments. A technical problem prevented the flight, but because earlier A-series rocket launches had succeeded, Lindbergh persuaded Goddard to donate a complete A-series rocket to the Smithsonian.

The period between the world wars saw the advent of small airplanes built for a civilian market. The **Staggerwing C17L**—offsetting the wings increased pilot visibility—made by the Beech Aircraft Company of Wichita, Kansas, was designed as a high-speed, comfortable business airplane. The Staggerwing in the Smithsonian, serial number 93, was manufactured on July 3, 1936, and listed for $10,260. Its original owner was Edwin Aldrin (father of astronaut Buzz Aldrin), a former army test pilot who worked for Standard Oil.

Perhaps the most accomplished pilot ever to fly a Staggerwing was **Jacqueline "Jackie" Cochran**, the first director of the Women Airforce Service Pilots (WASP) during World War II. Before the war, Cochran set two women's speed records in her Beech D-17W Staggerwing, followed by a victory in the prestigious Bendix Trophy Race in a Seversky P-35 in 1938. At the time of her death in 1980, she held more speed, altitude, and distance records than any other male or female pilot in aviation history.

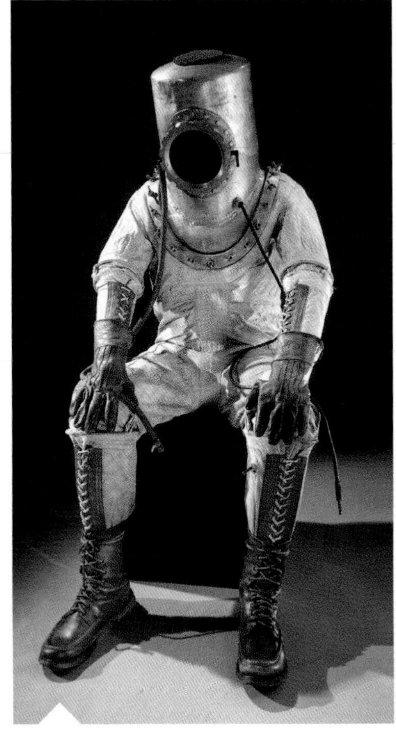

In 1933, flying a Lockheed Vega equipped with a Sperry autopilot, Wiley Post was the first person to make a solo round-the-world flight. The *Winnie Mae*, now in the Smithsonian, made the trip in seven days, eighteen hours. Two years later, Post wore this B.F. Goodrich **pressure suit**—the world's first—during a high-altitude flight in *Winnie Mae* from California to Cleveland.

Launched from Rapid City, South Dakota, on November 11, 1935, the **Explorer II balloon gondola** carried US Army Air Corps Captains Albert Stevens and Orvil Anderson to a then world record altitude of 72,395 feet, or nearly fourteen miles—high enough to photograph the curvature of the Earth for the first time.

Amelia Earhart would be remembered as one of history's most famous aviators even if she hadn't mysteriously disappeared over the Pacific during an attempted round-the-world flight in 1937. She began breaking aviation records within a year of her first flying lesson in 1921. Many were set in this Lockheed 5B Vega, which she called her *Little Red Bus*. In 1932, she flew it alone across the Atlantic Ocean, then took it nonstop across the United States—both firsts for a woman. The Vega wasn't up to a round-the-world trip, however. For that flight, she piloted a Lockheed Model 10 Electra.

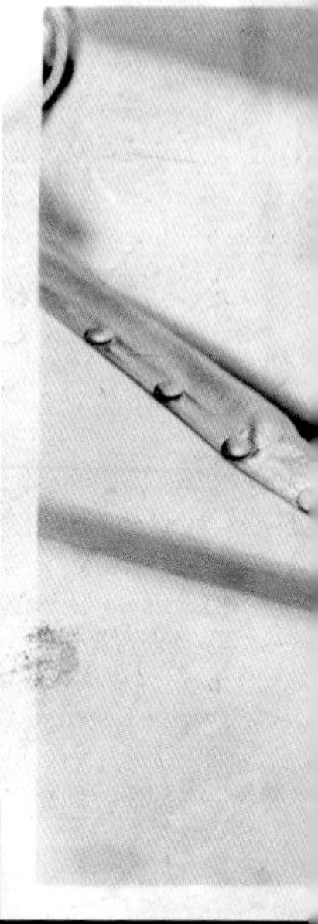

Earhart's **leather coat,** with tweed wool lining, was standard attire for military and civilian pilots, as it provided warmth and protection from the elements.

Earhart's husband, publisher George Putnam, commissioned this **teakwood chest** to hold her many trophies and plaques and other memorabilia.

Earhart wore these **flight goggles** during the 1929 Women's National Air Derby, one of the first air races for women.

One of the most successful airliners in history, the **Douglas DC-3** popularized air travel with its comfort, speed, and safety. Introduced in 1935, it was the first airliner to profit from carrying passengers only. More than thirteen thousand DC-3s were produced, many of which are still flying. The museum's DC-3 flew more than 56,700 hours with Eastern Air Lines. Its last commercial flight was on October 12, 1952, on a trip from San Salvador to Miami.

The twenty-one-seat DC-3 felt safe and sturdy by the standards of the day, partly due to its strong, multiple-spar wing and its all-metal construction.

Introduced in 1938, the **Boeing 307 Stratoliner** was the first airliner with a pressurized fuselage. It could carry thirty-three passengers and cruise at twenty thousand feet—above most bad weather, which made for a faster and smoother ride. The Stratoliner's wide fuselage was fitted with sleeper berths and reclining seats. Only ten were built, and just two airlines, TWA and Pan American, entered them in service. The airplane displayed in the Smithsonian was flown by Pan American as the *Clipper Flying Cloud.*

Pan American president Juan Trippe may have done more than any other person to popularize air travel to all parts of the world. Trippe kept this nineteenth-century **globe** in his New York office and would stretch a string between two points to calculate the distance and time it would take for his airliners to fly between them. The globe was featured prominently in many publicity photos of Trippe and became part of Pan Am's public image.

WORLD WAR II

Aircraft had played an important supporting role in World War I, but when war erupted again a generation later, they were essential. Workhorse transport planes like the Douglas C-47 carried cargo and troops. Bombers like the B-17 and B-26 flew ever higher and farther, raining destruction on cities and factories, with nimble fighters as escorts. Naval battles were fought by dive-bombers and other attack aircraft, while the ships remained out of sight of one another. And the war ended with the blinding flash of atomic bombs dropped from B-29s.

(Artwork: *Study for Fortresses Under Fire*, Keith Ferris, 1976)

With its combination of speed, range, maneuverability, and firepower, many people consider the North American **P-51 Mustang** the best fighter of World War II. Fighter escorts protected long-range bombers from aerial attack and were used for strafing and reconnaissance. Although the museum's P-51D Mustang never saw combat, it is painted as *Willit Run?*, which flew with the 351st Fighter Squadron, 353rd Fighter Group, Eighth Air Force.

Among the best-known Mustang squadrons were those of the 332d Fighter Group, some of whose pilots are shown here in Ramitelli, Italy, in August 1944. Known as the **Tuskegee Airmen**, after the institute in Alabama where they trained, the unit was composed mostly of African Americans.

The cockpit of the museum's **Supermarine Spitfire Mk VII** shows the pilot's cramped workspace. Alongside the **Hawker Hurricane** (below), another legendary fighter represented in the Smithsonian collection, Spitfires defended against the German Luftwaffe during the Battle of Britain.

Introduced in the 1930s, the **Messerschmitt Bf 109** single-seat fighter was mass-produced during the war and became a mainstay of Germany's air force. This Bf 109 G-6/R3 was transferred to the Smithsonian in 1948. When plans for a new museum building on the Washington Mall became definite in the mid-1970s, it was one of the first aircraft restored for exhibition.

This **glove** belonged to German Bf 109 pilot Günther Rall, one of the Luftwaffe's top aces. Rall lost his thumb in an aerial battle on May 12, 1944, and was forced to bail out. He survived to become a high-ranking officer in the West German air force during the Cold War. His diary is also in the museum.

This **Martin B-26B Marauder** nicknamed *Flak-Bait* flew 202 bombing missions during the war, more than any other American aircraft—including three raids on D-Day alone. *Flak-Bait*'s crews dropped bombs on cities, airfields, roads, bridges, supply depots, and practically every other kind of target. By war's end, ground crews had repaired some one thousand holes caused by Nazi shrapnel, cannon shells, and machine gun bullets, earning the durable twin-engine bomber its name.

Scrawled on *Flak-Bait*'s fuselage are hundreds of **signatures** from maintenance workers and others who signed it at the end of the war. This one, by Wiley L. Flesher of Baltimore, Maryland, is in the top of the B in *Flak-Bait*.

These **photos**, from the museum's Aerial Photographic Reconnaissance (Batchelder) Collection, show Germans preparing beach defenses on the Normandy coast a month before the D-Day invasion on June 6, 1944. The pictures were taken—at great risk—by low-flying Allied recon aircraft. Note the Germans scrambling and lying prone on the beach as the airplanes pass overhead.

Martin B-26 crews typically included a pilot, copilot, bombardier, tail gunner, radio operator, and turret gunner. This particular ***Flak-Bait* crew** flew together eleven times between December 1944 and March 1945, but each man flew many more times with other crews.

Developed in secret at the Peenemünde test site on the Baltic Sea and built by concentration-camp prisoners, Germany's **V-2** (Vergeltungswaffe Zwei, or "Vengeance Weapon Two") was the world's first large-scale liquid-propellant rocket vehicle and the first long-range ballistic missile. Beginning in late 1944, thousands of V-2s were fired on London, Antwerp, and other cities. After the war, captured V-2 hardware and engineers, including Wernher von Braun, helped jump-start the early American missile program. The US Air Force officially transferred this V-2 to the Smithsonian on May 1, 1949.

A **V-2** rises from the Peenemünde range during a test launch in summer 1943.

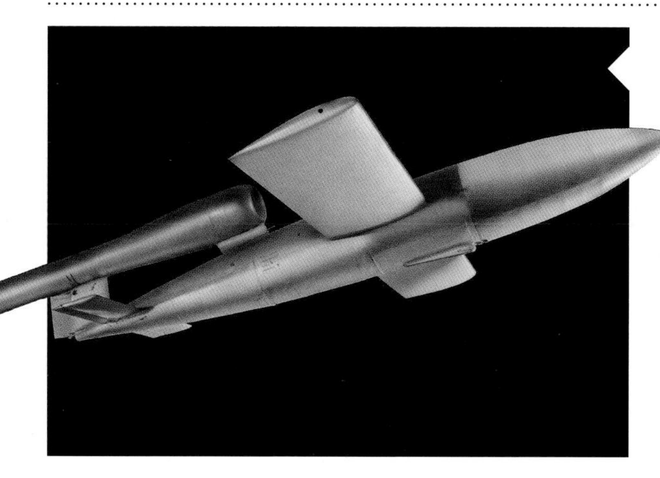

The German-made **V-1** was the world's first operational cruise missile. Powered by a simple but noisy pulse-jet engine that earned it the nicknames *Buzz Bomb* and *Doodle Bug*, more than twenty thousand were launched at British and continental European targets from June 1944 to March 1945. The V-1 had a range of about 150 miles but was very inaccurate. The US Air Force transferred this one to the Smithsonian in 1949.

Posters, advertisements, and other mass media contributed to the war effort. This ad from the Smithsonian's collection portrays the Douglas Dauntless dive-bomber, hero of the Battle of Midway.

The **Soviet Ilyushin Il-2 Shturmovik** was the preeminent ground-attack aircraft of World War II, devastating the German army on the Eastern Front. Flying many missions unescorted, Shturmoviks earned a reputation for strength and durability. A total of 36,163 were built, more than any other combat aircraft of the war.

The museum's art collection includes a series of portraits taken by photographer Anne Noggle—a former pilot herself—of American and Soviet women who had served during World War II, including munitions mechanic **Yekaterina Chujkova**.

In 2009, the Congressional Gold Medal was awarded to the hundreds of members of the **Women Airforce Service Pilots (WASP)**, civilian employees who tested aircraft, trained other pilots, and ferried airplanes for the military during World War II.

WASP **Betty Jane Bachman** waves from the cockpit of a Curtiss P-40 Warhawk.

The **Boeing B-29 Superfortress** was the most sophisticated propeller-driven bomber of World War II and the first bomber with pressurized compartments. On August 6, 1945, this Martin-built B-29-45-MO, named *Enola Gay* after pilot Paul Tibbets's mother, dropped the first atomic weapon on Hiroshima, Japan. Restoration of the aircraft—the largest such project ever undertaken at the museum—took nineteen years and approximately three hundred thousand person-hours before *Enola Gay* went on display at the Steven F. Udvar-Hazy Center in Virginia in 2003.

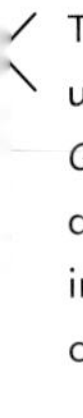

The **Norden bombsight** used on the *Enola Gay* for the first atom bomb drop is in the Smithsonian collection.

From this **tail gunner's station**, Technical Sergeant George Robert "Bob" Caron witnessed (and photographed) the mushroom cloud rising over Hiroshima.

The museum staff meticulously restored the **cockpit** of the B-29 *Enola Gay*.

JET AGE MEETS SPACE AGE

Just as the first great war introduced innovations to aviation, the second pushed technology even further. The first jet-powered airplanes flew before the war's end. Cold War tensions led the leading industrial nations to fly higher and faster, to spy on one another, and to increase their bombing capability. And missiles turned into launch vehicles that sent the first objects into orbit. The Space Age had arrived.

(Artwork: *Bell X-1 Rocket Powered Aircraft*, Robert C. Korta, ca. 1960)

Although there was never a literal “sound barrier,” airplane designers struggled with drag and turbulence affecting aircraft as they approached the speed of sound. On October 14, 1947, on its fiftieth flight, Bell Aircraft’s rocket engine–powered **X-1** became the first airplane to go supersonic. Piloted by US Air Force Captain Charles E. “Chuck” Yeager, X-1 #46-062 reached a speed of seven hundred miles per hour, Mach 1.06, over the Mojave Desert. The historic aircraft, nicknamed *Glamorous Glennis* for Yeager’s wife, was delivered to the museum in 1950.

Yeager was one of several test pilots to fly the X-1. At his urging, the museum painted the airplane orange to match its appearance on the day of his first Mach 1 flight.

The Bell X-1 cockpit display included a **Machmeter**.

America's first swept-wing jet fighter, the **North American F-86 Sabre**, dueled with Soviet-built MiG-15s during the Korean War—some of the first jet dogfights. This F-86A flew most of its missions from Kimpo Air Base near Seoul and bears the markings of the Fourth Fighter Wing.

The jet age prompted research into how much acceleration (measured in g-force) pilots—including those ejecting from high-speed aircraft—could stand. Air Force Lieutenant Colonel John Stapp rode this **Sonic Wind 1 rocket sled** three times during tests at Holloman Air Force Base in New Mexico. On December 10, 1954, he made his last and most notable run, reaching a speed of 632 miles per hour in five seconds. Stapp endured a force of more than 40 g before he stopped, the highest any human had withstood up to that time.

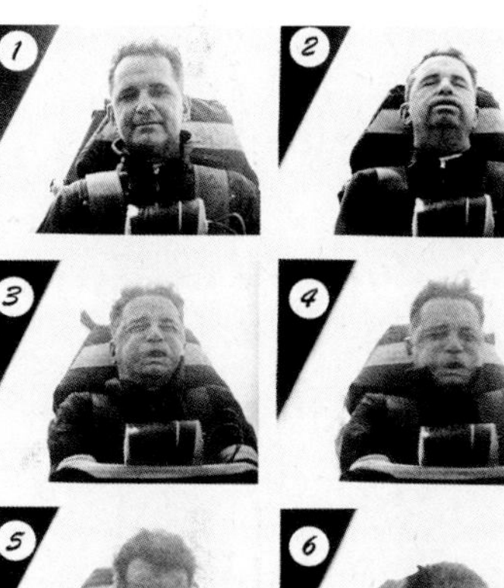

Stapp experienced a force of 22 g during a test in June 1954.

The **Bell 47** was the first civilian helicopter to be certified. Popularized first in the 1950s television series *Whirlybirds* and later in the series *M*A*S*H*, it served in Korea and Vietnam. Over a period of forty years, this Bell 47B, the thirty-sixth built, did many jobs, including news gathering, crop dusting, and aerial photography.

A pilot from the age of twelve, **Betty Skelton** paved the way for women in sport aviation. The first woman to perform the inverted ribbon cut at air shows, she won the International Feminine Aerobatic Championships in 1948, 1949, and 1950 and set seventeen records for flying and driving race cars.

In addition to her trademark **Pitts-Special S-1C,** *Little Stinker,* Skelton donated to the museum a collection of news clippings, pamphlets, magazines, photographs, and scrapbooks covering the span of her career.

One of many small airplanes marketed to private pilots by companies like Cessna, Piper, and Beech, the **Cessna 180** was introduced as a rugged, high-wing utility aircraft in 1952. The museum's Cessna, *Spirit of Columbus*, made news in 1964 when Jerrie Mock became the first woman pilot to complete a solo flight around the world.

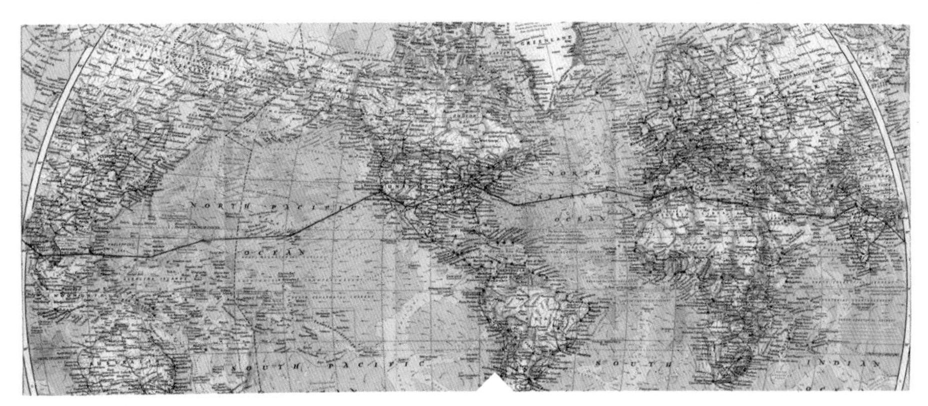

Mock departed Columbus, Ohio, and flew an easterly course, making fourteen stops in twenty-nine days, eleven hours. This **hand-marked map** traces her route.

The first artificial satellite, **Sputnik 1** (which means "traveling companion") was launched on October 4, 1957, and remained in orbit for ninety-two days. The museum's replica is one of many manufactured by the same Soviet factory that made the original.

In the late 1950s, the United States and the Soviet Union raced to develop ever-more-powerful ballistic missiles that could hurl bombs across the world. Modified, sometimes with additional upper stages, those same missiles could also launch small objects and then increasingly larger ones into orbit—the Earth's first artificial satellites. The new spacecraft, as they came to be called, were soon put to all manner of uses, from science to communications and weather observation.

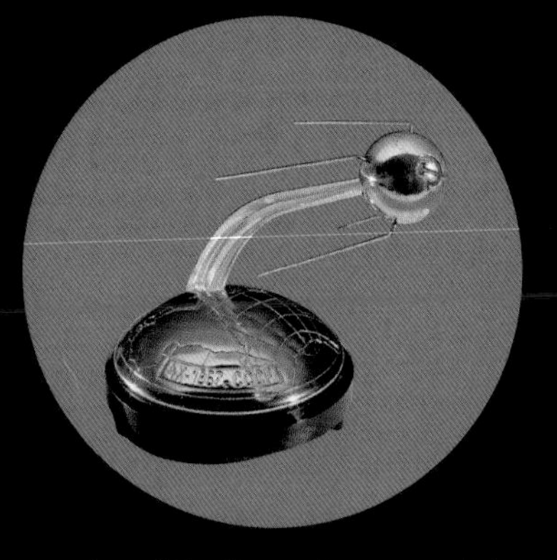

The opening of the Space Age was celebrated in pop culture around the world. A private collector who purchased this **Sputnik music box** while on a trip to the Soviet Union in 1964 donated it to the museum in 1985.

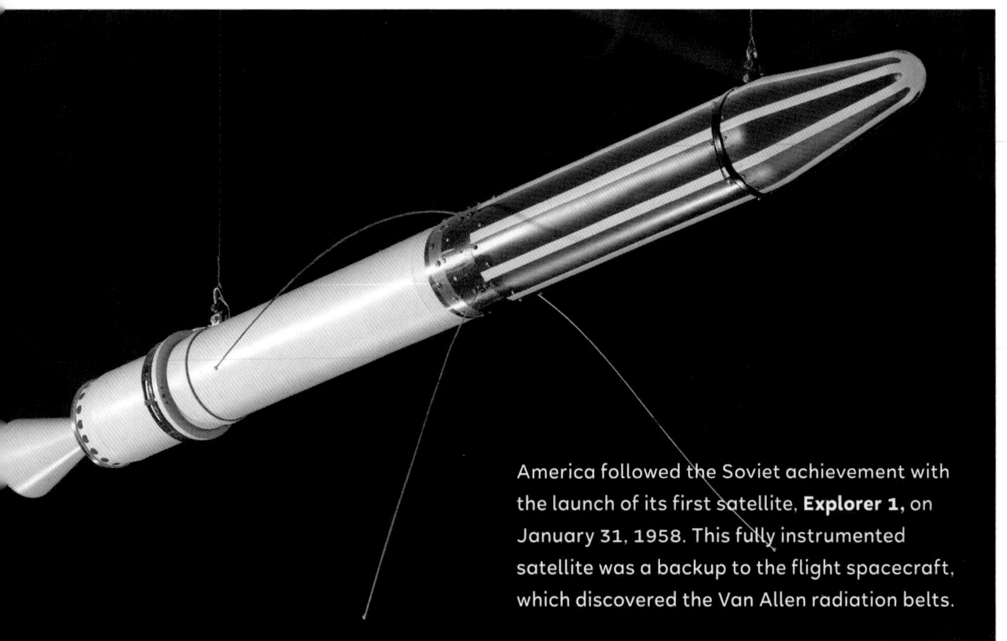

America followed the Soviet achievement with the launch of its first satellite, **Explorer 1,** on January 31, 1958. This fully instrumented satellite was a backup to the flight spacecraft, which discovered the Van Allen radiation belts.

TIROS 1 (Television Infrared Observation Satellite), launched by the United States in April 1960, was the world's first weather satellite. Over its operational lifetime of two and a half months, the satellite transmitted thousands of **TV images** of cloud patterns to ground stations.

This **engineering model** of the magnetic-tape data recorder on board TIROS 1 was assembled from spare components.

North American's rocket-powered **X-15** research aircraft first flew in 1959, bridging the gap between aviation and spaceflight. Dropped from a B-52, it then used its own propulsion to climb to the edge of space, and was the first winged aircraft to attain hypersonic velocities up to Mach 6. Three X-15s were built. The one in the Smithsonian, X-15 #1, was flown by test pilot Neil Armstrong on the last of his seven X-15 flights in July 1962, shortly before he was chosen as a NASA astronaut.

On May 5, 1961, Alan Shepard became the first American in space. Launched on top of a Redstone rocket, his black-painted Mercury capsule reached an altitude of 116 miles, almost twice as high as the X-15. Five months later, NASA gave Shepard's *Freedom 7* **Mercury spacecraft** to the Smithsonian.

Shepard's **Mercury spacesuit** was a close-fitting, two-layer, full-pressure suit developed by the B.F. Goodrich Company, which based it on its Mark IV pressure suit used by the US Navy.

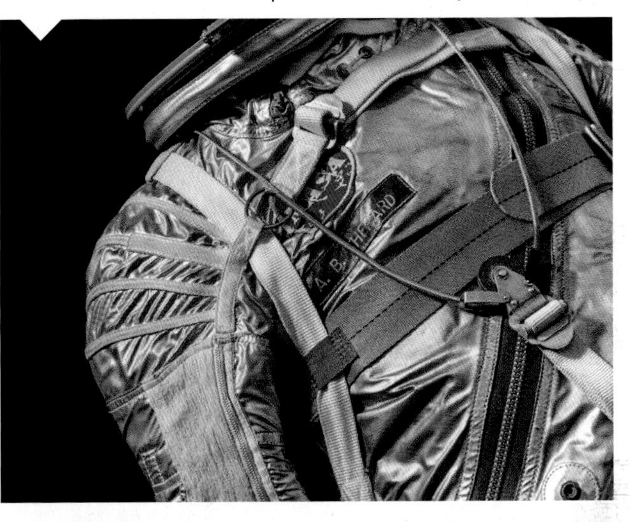

Although Shepard's flight lasted only fifteen minutes, it proved that the **Mercury capsule** and ocean recovery worked. Shepard is shown here shortly after spacecraft and astronaut were helicoptered separately to a waiting aircraft carrier.

J37
J20

In this **Mercury capsule** on February 20, 1962, John H. Glenn Jr. (pictured during the flight) became the first American to orbit the Earth. NASA transferred *Friendship 7* to the Smithsonian a year later. In 1998, just before retiring from the US Senate at the age of seventy-seven, Glenn made his second trip into space, aboard Space Shuttle *Discovery*.

Before his Mercury-Atlas 6 flight, Glenn stopped at a drug store and bought this **Ansco camera** to go along with a Leica supplied by NASA. Because his helmet made it hard to see through the built-in viewfinder, NASA engineers attached a larger viewfinder on top, plus a pistol grip to simplify handling in microgravity.

Glenn took this **photo** from inside *Friendship 7* using his store-bought Ansco camera.

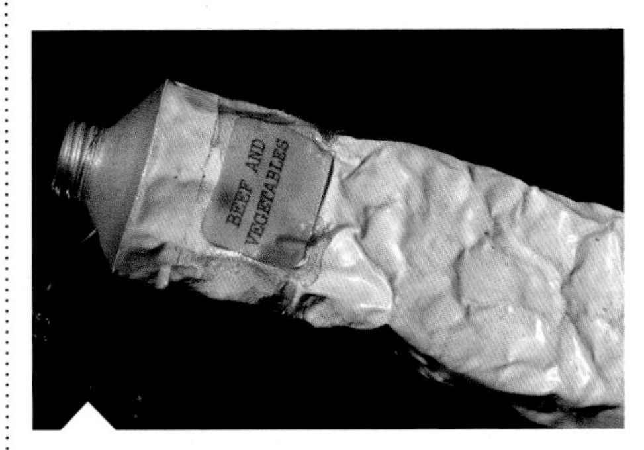

This squeezable **tube of pureed food**—no one was sure at the time how difficult eating would be in zero gravity—was issued to Glenn for his five-hour *Friendship 7* flight.

In the 1960s, the **Boeing 707** was synonymous with stylish airline travel. Before it came the 367-80, or Dash 80, a quadjet prototype first flown by Boeing in 1954 to demonstrate the advantages of jet propulsion for commercial aviation. Only one was built. In 1972, the distinctive brown-and-yellow jet, forerunner of the 707 as well as the KC-135 military tanker, was donated to the Smithsonian.

Pan American began Boeing 707 **passenger service** from New York to Paris in October 1958.

The first **Learjet** aircraft, the **Model 23**, pioneered the new field of business jet aviation. A Learjet became a kind of status symbol and was a favorite of celebrities, including Elvis Presley, Frank Sinatra, and James Brown, the first African American to own a private jet.

The museum's **Learjet**, the second prototype built, flew 1,127 hours and 864 flights as a test aircraft.

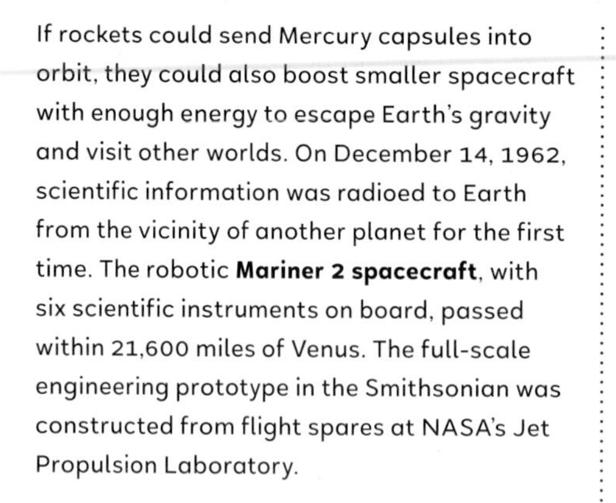

If rockets could send Mercury capsules into orbit, they could also boost smaller spacecraft with enough energy to escape Earth's gravity and visit other worlds. On December 14, 1962, scientific information was radioed to Earth from the vicinity of another planet for the first time. The robotic **Mariner 2 spacecraft**, with six scientific instruments on board, passed within 21,600 miles of Venus. The full-scale engineering prototype in the Smithsonian was constructed from flight spares at NASA's Jet Propulsion Laboratory.

Mariner 2, the world's first successful interplanetary spacecraft, beamed back **data** on Venus and its atmosphere to eagerly waiting scientists at the Jet Propulsion Laboratory in Pasadena, California.

Despite their scientific-sounding cover name, the **Galactic Radiation and Background (GRAB)** spacecraft were the world's first successful spy satellites, built to eavesdrop on radar signals from America's Cold War adversaries. Five GRAB satellites were launched from June 1960 to April 1962. The Naval Research Laboratory donated this backup spacecraft in 2002.

Launched on July 10, 1962, **Telstar 1**, developed by the American Telephone and Telegraph Company (AT&T), was the world's first active communications satellite. This backup spacecraft was transferred to the museum in 2006.

Introduced in the mid-1960s, the Lockheed **SR-71 Blackbird** spy plane is still the fastest piloted jet aircraft ever built, with a cruising speed above Mach 3. The museum's Blackbird (61-7972) logged about twenty-eight hundred hours of flight time during twenty-four years of service with the US Air Force. During its final flight on March 6, 1990, Lieutenant Colonels Ed Yielding and Joseph Vida set a speed record by flying from Los Angeles to Washington, DC, in one hour, four minutes, and twenty seconds, averaging 2,124 miles per hour. For more than a decade before it went on display at the Steven F. Udvar-Hazy Center in 2003, the Smithsonian's SR-71 was stored in a custom hangar built for its protection in a secured area at nearby Dulles Airport.

Two **SR-71s**, including this one shown flying over the California desert in 1997, were loaned to NASA in the 1990s for high-speed flight research.

This **flying scarf** was worn by members of an SR-71 Strategic Reconnaissance Squadron. *Habu*, a nickname for the plane, comes from a snake native to Okinawa.

The 1960s saw the United States embroiled in another war halfway across the world in the jungles of Southeast Asia. American aircraft of all types, from giant B-52 bombers to F-4 Phantom fighters to small Cessna Bird Dog reconnaissance aircraft, ruled the sky over Vietnam, while stalwart **Huey helicopters** did practically every job on the battlefield, from carrying troops into combat to evacuating the wounded.

The **Bell UH-1 Iroquois helicopter**, commonly known as the Huey, became an indelible symbol of the Vietnam War. From 1966 to 1970, this UH-1 compiled a distinguished combat record with four units, including the 229th Assault Helicopter Battalion of the First Cavalry and the 118th and 128th Assault Helicopter Companies. Numerous patches on the helicopter's skin attest to the ferocity of missions flown while operating as a "smoke ship," laying down smoke screens for air assault operations with the Eleventh Combat Aviation Battalion.

The US Air Force, Navy, and Marine Corps all flew the **McDonnell Douglas multi-role F-4 Phantom II**. In this aircraft—which was at the time a Navy F-4J—Commander S. C. Flynn and his radar intercept officer, Lieutenant W. H. John, spotted three enemy MiG fighters off the coast of Vietnam on June 21, 1972, and shot down one with a Sidewinder air-to-air missile. This Phantom also flew combat air patrols and bombing missions during the Linebacker II bombing campaign that same year.

Hundreds of captured American servicemen, many of them downed pilots, spent time in Vietnamese prison camps like Hỏa Lò, better known as the Hanoi Hilton. This **striped top** was worn by one of the prisoners.

THE MOON AND BEYOND

Three weeks after Alan Shepard's first US spaceflight in May 1961, President John F. Kennedy set a bold goal: land astronauts on the Moon and return them to Earth before the decade is out. That single-minded mission was accomplished in stages with the Mercury, Gemini, and Apollo flight programs. Robotic spacecraft were sent to the Moon ahead of the landings to scout the lunar environment. At the same time, NASA and the Soviet Union took the first steps toward robotic exploration of the solar system.

(Artwork: *Study for Lunar Landscape Mural*, Chesley Bonestell, ca. 1955)

Before astronauts could land on the Moon, spacecraft were dispatched to study it up close, at much higher resolutions than telescopes on Earth could see. Starting in 1964, three Ranger spacecraft crashed cameras into the Moon, kamikaze-style, sending ever-closer pictures right up to the moment of impact. Five Lunar Orbiters scouted Apollo landing sites from above, while five Surveyors touched down on the lunar surface—the first American landings on another world.

The final Ranger spacecraft, **Ranger 9**, crashed into the Moon on March 24, 1965. This sequence of TV images documents its descent toward Alphonsus Crater. The last picture, received just before impact, is at lower right.

Each successful Ranger spacecraft had **six television cameras** with different exposure times, fields of view, lenses, and scan rates. The full-size spacecraft on exhibit in the Smithsonian, made of parts from test vehicles, is a replica of the last Rangers in the series.

This **first view of the Earth** from the vicinity of the Moon was taken by Lunar Orbiter 1 on August 23, 1966.

Five **Lunar Orbiters** mapped 99 percent of the Moon at resolutions down to one meter. The museum's engineering mockup shows the spacecraft's square solar panels, antenna boom, fuel tanks, and rocket engine for velocity control.

Five Surveyors landed on the Moon between June 1966 and January 1968. In November 1969, Apollo 12 landed close enough to **Surveyor 3** for astronauts Charles "Pete" Conrad and Alan Bean to pay a visit. Note the arm-like scoop extending to the right in front of Conrad.

The Surveyor spacecraft on display in the museum was used for prelaunch thermal control tests on Earth. It was later reconfigured to represent the later-model Surveyors that carried a scoop and claw-like surface sampler to the Moon. The Smithsonian collection also includes this engineering model of the **extendable soil mechanics surface scoop** (SMSS).

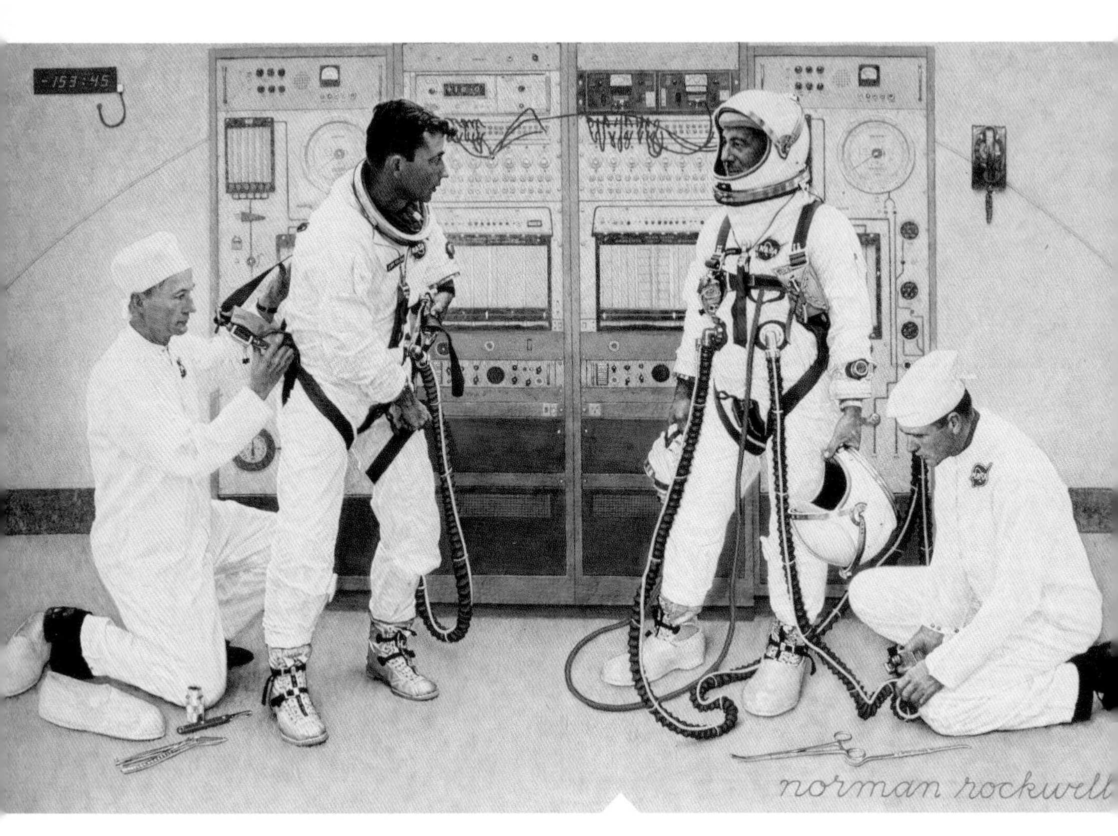

After Mercury proved that astronauts could survive orbital flight and that its spacecraft design was sound, NASA advanced to two-man flights with Gemini. Over the course of ten missions, this crucial test program demonstrated spacecraft rendezvous, the first spacewalks, and extended-duration flights lasting two weeks—long enough to show that trips to the Moon and back were feasible. As in Mercury, the astronauts and their capsules splashed down in the ocean at mission's end.

Beginning in 1962, NASA commissioned artists, such as Robert Rauschenberg and Jamie Wyeth, to document the early space program. This 1965 oil painting by **Norman Rockwell**, now in the Smithsonian collection, shows astronauts John Young and Gus Grissom suiting up for the first Gemini flight in March 1965. NASA loaned Rockwell a Gemini spacesuit so he could depict the scene as accurately as possible.

Ed White made the **first US spacewalk** on June 3, 1965, during the Gemini IV mission. Tethered to the spacecraft, he stayed outside for twenty-three minutes while crewmate Jim McDivitt took pictures.

The **gloves worn by White** during his historic spacewalk had two layers: an internal rubber (neoprene) pressure bladder and a nylon upper cover layer. They were designed to prevent objects from slipping out of White's hands as he floated in weightlessness.

Frank Borman and James Lovell were launched into orbit aboard this spacecraft, **Gemini VII**, on December 4, 1965. They stayed in space for fourteen days, a record that would stand until 1970. Gemini VII also served as the target vehicle for the world's first spacecraft rendezvous, with Gemini VI-A, on December 15. The Smithsonian exhibits two Gemini spacecraft, Gemini IV and Gemini VII.

Because their Gemini VI mission happened shortly before Christmas 1965, Wally Schirra and Tom Stafford radioed a holiday message down to the ground. While Schirra blew "Jingle Bells" on a harmonica, Stafford played these **bells**. The astronauts donated both instruments to the Smithsonian.

Landing a dozen people on the Moon between July 1969 and December 1972 and returning them safely to Earth required thousands of technical innovations, from rocketry and spacecraft design to medical sensors and food packaging. For more than fifty years after its debut, the giant Saturn V booster that sent the Apollo spacecraft on their way ranked as the most powerful ever built. The Smithsonian holds artifacts of all types from what many consider the greatest engineering project of the twentieth century.

With each of its five F-1 first-stage engines producing 1.5 million pounds of thrust, a **Saturn V rocket** lifts off from Cape Canaveral, Florida, on July 16, 1969, with Apollo 11 astronauts Neil Armstrong, Buzz Aldrin, and Michael Collins on board—the start of the first lunar landing mission. Along with several F-1s used for testing, the Smithsonian has pieces of the actual engines from Apollo 11, which were recovered from the ocean floor in 2013.

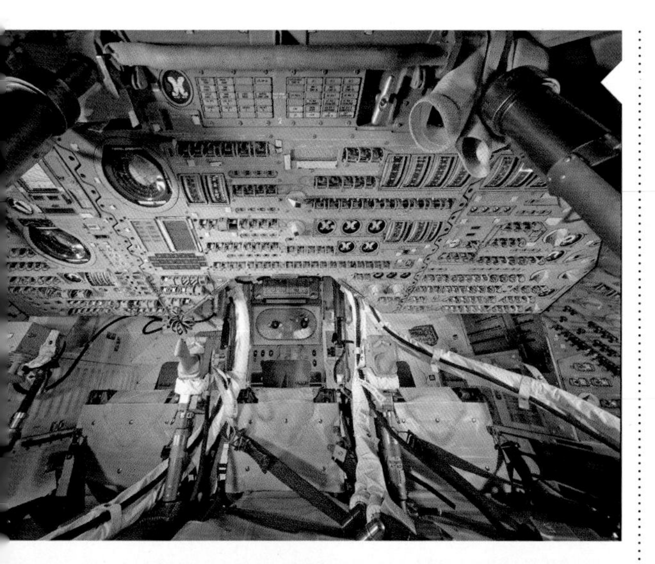

The only one of the Apollo spacecraft's three parts that returned to Earth was the command module. This **interior view** shows the Apollo 11 *Columbia*, now on display in the Smithsonian.

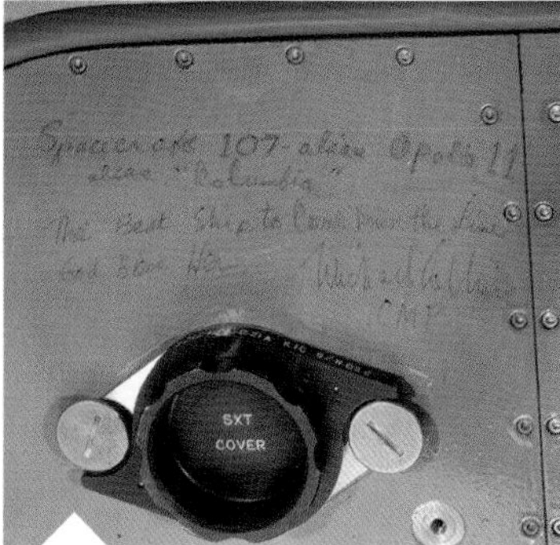

Shortly after his return to Earth, astronaut Michael Collins went back inside *Columbia* to inscribe this **message** on one of the equipment bay panels.

This is the **spacesuit** worn by Neil Armstrong on the Moon. The Type A-7L suit underwent an extensive conservation process, which was funded by thousands of public donations raised through a Kickstarter campaign in 2015.

This film magazine—Magazine S— from the 70mm Hasselblad camera used by Armstrong and Aldrin on the lunar surface, contained some of Apollo 11's most famous images, including this **sequence** of Aldrin descending the ladder of the lunar module. The camera itself is still on the Moon.

In the unlikely event they introduced potentially dangerous organisms from the Moon, the Apollo 11 astronauts (from left: Collins, Aldrin, and Armstrong) lived in isolation for three weeks. This **mobile quarantine facility**, which transported them from the recovery carrier to the Lunar Receiving Lab in Houston, is now on display at the Steven F. Udvar-Hazy Center.

The **Apollo lunar module (LM)** that ferried astronauts from lunar orbit to the Moon's surface and back had two main parts. The upper ascent stage (above the gold-colored foil) consisted of a pressurized crew compartment and a rocket engine for lifting off from the Moon after the astronauts finished exploring. The lower descent stage remained on the Moon. The Smithsonian's lunar module, LM-2, was originally built for an Earth-orbit test flight that was canceled. All the LMs were slightly different; this one was modified to appear like the Apollo 11 LM *Eagle*.

The **first telescope used on another world** was the lunar surface camera designed by George Carruthers of the Naval Research Laboratory, one of the few African American scientists to work on Apollo. Astronaut Charles Duke is shown here with the deployed telescope during the Apollo 16 mission. The Smithsonian has a qualification test unit with the actual film cartridge brought back from the Moon.

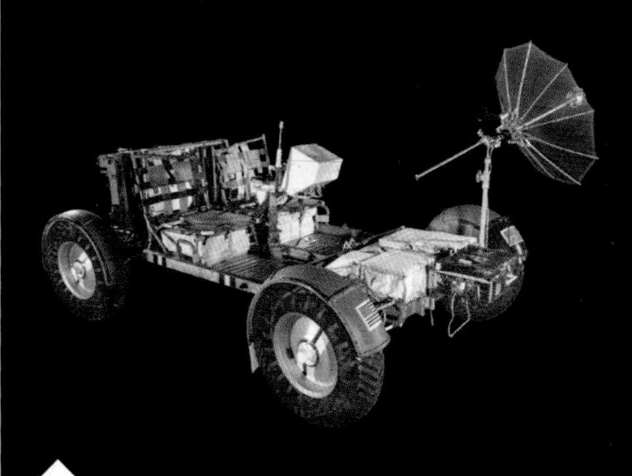

The **lunar roving vehicle (LRV)** was a battery-powered "dune buggy" driven by astronauts on the Moon on Apollo missions 15, 16, and 17. This qualification test unit, used for preflight stress testing, is a very close replica of the vehicles that flew.

The **lunar "touchstone"** is one of the museum's most popular exhibits—a rare piece of the Moon that visitors can touch for themselves. Apollo 17 astronauts brought this chunk of iron-rich volcanic rock back from the Taurus-Littrow valley in 1972, and NASA subsequently loaned it to the Smithsonian.

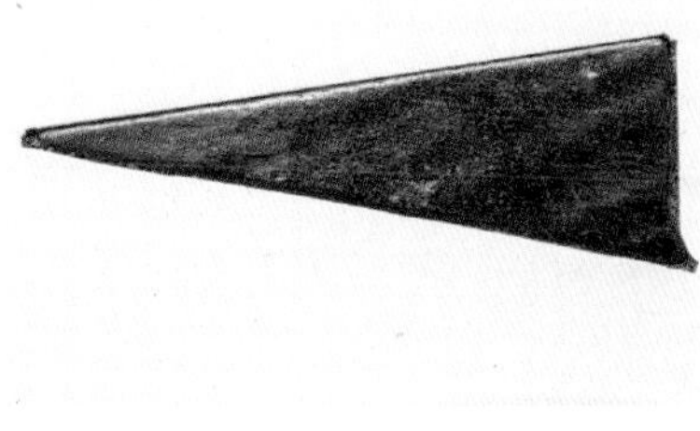

In the 1960s and 1970s, US and Soviet probes zoomed past Mars and Venus, taking pictures and collecting data. In 1971, NASA's Mariner 9 arrived at Mars, becoming the first spacecraft to orbit another planet. Finally, in 1976, two Viking landers touched down on the Martian surface carrying a suite of biology and chemistry experiments—the first science investigations on another planet.

On display in the Smithsonian is the proof test article of the **Viking Mars lander**. NASA's Viking program sent two orbiters and two landers to Mars. Viking 1 came down in the Chryse Planitia region on July 20, 1976, and Viking 2 landed in Utopia Planitia six weeks later. Both landers operated for years and are still on the Martian surface. The dish-like antenna was for sending data back to Earth.

The **Viking 2 lander** rests on Martian soil at the Utopia landing site. The flag and color chart on the spacecraft helped scientists calibrate the true color of the planet's surface and sky, which appears reddish due to suspended dust.

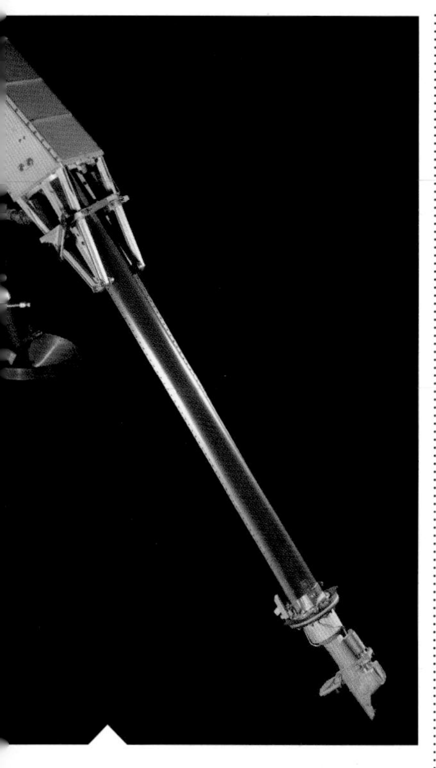

The **long digging arm** on the Viking lander, like this one on the museum's spacecraft, collected soil for experiments.

The Soviet Union's planetary exploration program saw its greatest success at Venus. In 1984, the Soviet-led multinational Vega 1 and Vega 2 spacecraft flew past the planet, dropped balloons and landers, and then went on to pass through the tail of Comet Halley. On display in the museum are engineering models of the **Vega spacecraft** bus and a balloon that took measurements in the Venusian atmosphere.

Soviet Venera landers also returned photos of the rock-strewn Venusian surface, including this **color panorama** taken by Venera 14 in 1982.

Skylab was the first US space station, built largely from hardware developed for the Apollo Moon program. Following the launch of the empty station on Skylab 1, three separate crews of three astronauts each spent weeks at a time in 1973 and 1974 learning to live and conduct experiments inside the station for extended periods in orbit. The longest mission, Skylab 4, lasted eighty-four days.

A TV camera in the station's wardroom transmitted video of the **Skylab 3 crew** having a meal.

Skylab's orbital workshop—modified from the third stage of the Saturn V rocket—included the crew's living quarters, work and storage areas, and research equipment. Two complete Skylab space stations were manufactured and equipped for flight, but only one was launched. In 1975, following the end of the Skylab program and NASA's shift to Space Shuttle development, this backup orbital workshop was transferred to the museum.

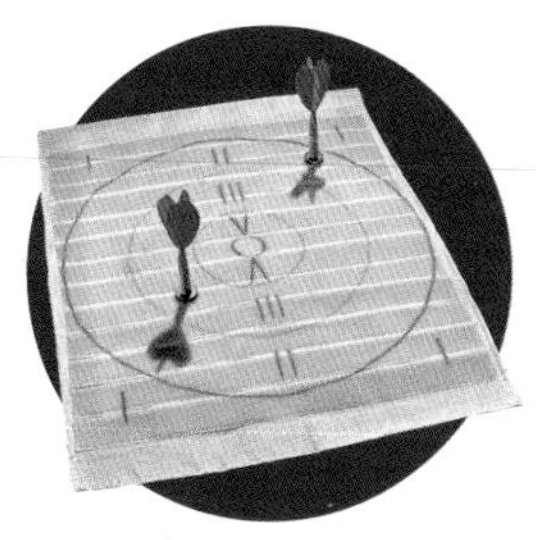

Designers of a long-term space station had to consider recreation as well as basic survival. This **darts game** is a replica of one flown aboard Skylab. For safe play in space, the sharp point was replaced with a Velcro patch. The Velcro Companies made this set and gave it to the museum.

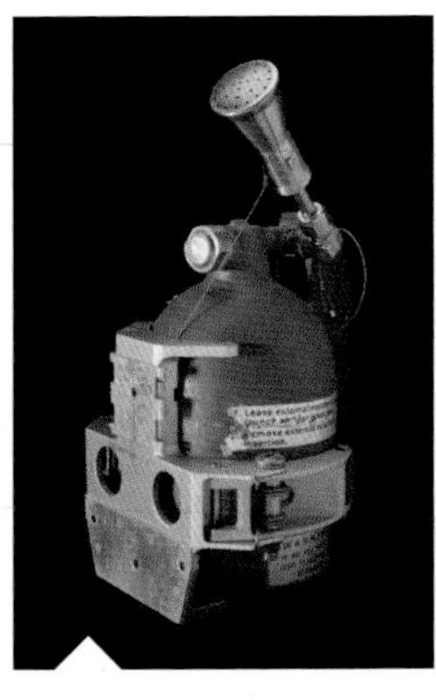

This **fire extinguisher** is a duplicate of the one used on the Skylab station in orbit.

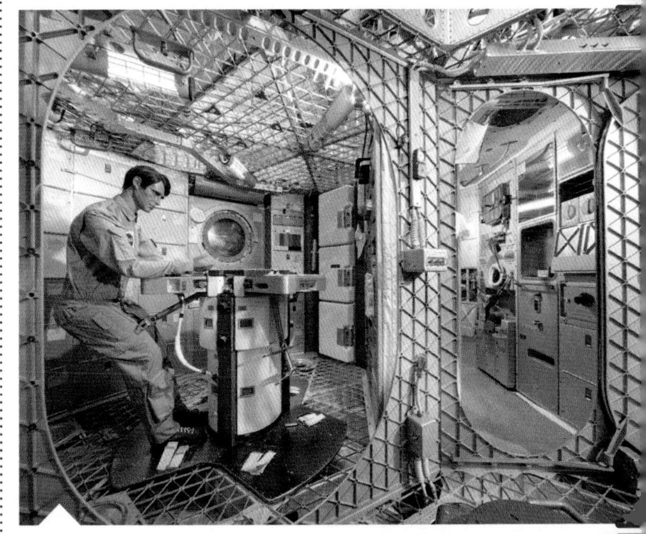

Museum visitors can walk through **Skylab's living quarters** and get a look inside.

PUSHING BOUNDARIES

In the last quarter of the twentieth century, aviation and space exploration advanced not in dramatic leaps and bounds, but rather with steady, incremental progress. Following the end of the Apollo Moon program, NASA turned to operating a new orbital transportation system—the Space Shuttle—and building a permanent space station. Planetary probes ventured into the outer reaches of the solar system for the first time. And passenger aircraft pushed the boundaries of size and speed with jumbo jets and the Concorde supersonic transport.

(Artwork: *Dale A. Gardner, Space Shuttle Mission 51-A*, George D. Guzzi Jr., 1985)

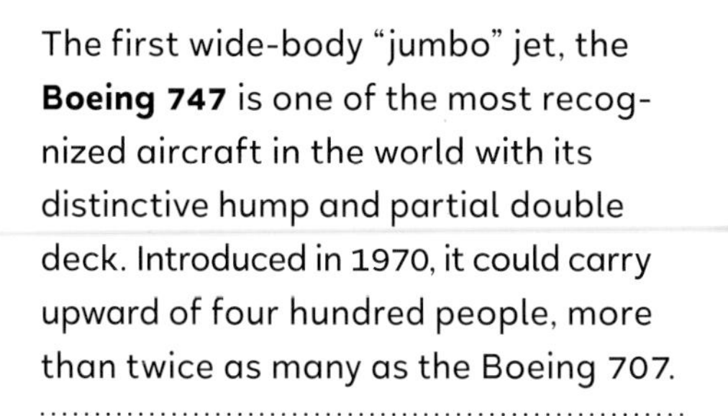

The first wide-body "jumbo" jet, the **Boeing 747** is one of the most recognized aircraft in the world with its distinctive hump and partial double deck. Introduced in 1970, it could carry upward of four hundred people, more than twice as many as the Boeing 707.

On display in the museum's National Mall building is this **nose section** from a Northwest Airlines Boeing 747-151, which visitors can enter from the second floor. If the entire 250-foot-long jet were on display, it would take up nearly half the length of the museum.

The 747 needed two aisles to comfortably seat ten passengers in each row.

Introducing:

CONTINENTAL

Diamond Head Lounge. First Class passengers: wind your way up here, then unwind—Hawaii style.

Kabuki Room. Oriental, delicate, tasteful. Coach as only Continental can do it. Beautifully.

Bougainville Room. Coach. Like all other rooms aboard, it offers living room luxury and spaciousness, with armchair comfort and broad aisles.

Kamehameha Room. Comfort is the keynote, First Class is the service, like no other First Class. It's fit for King Kamehameha himself and makes everybody feel very royal.

Micronesia Room. Economy with the flavor of the delightful Micronesian Islands. At the rear is our comfortable lounge. Only Continental offers Coach and Economy passengers a lounge of their own.

New. Beginning June 26 between Los Angeles and Hawaii. Beginning July 18 between Chicago, Los Angeles and Hawaii. Comfortable. Roomy seats in spacious rooms. Unique. Each room in the spirit of one part of the exciting Pacific. Proud. Manned by a proud flight crew, attentive hostesses and helpful Directors of Passenger Service. Throughout, we've combined the luxury of an ocean voyage with the convenience of flight. Take our Air Cruise to Hawaii. Book passage through Continental or your travel agent, our partner in getting things done for you.

The Proud Bird of the Pacific

The Proud Bird of the Pacific

Jumbo jets were particularly useful for long voyages to remote destinations like Hawaii.

The first supersonic airliner to enter service, the British-French Concorde flew thousands of passengers across the Atlantic at twice the speed of sound for more than twenty-five years, beginning in 1976. Although the Concorde was a technological marvel, the economics of supersonic air travel proved unsustainable. In 2003, Air France donated **Concorde F-BVFA** to the museum following the completion of its final flight. This aircraft was the first Air France Concorde to open service to Rio de Janeiro, Washington, DC, and New York and had flown 17,824 hours.

This view inside the cramped **cockpit** of the museum's Concorde shows space was tight for the crew of two pilots and one flight engineer.

Cruising at altitudes of up to sixty thousand feet, Concorde passengers saw a dark blue sky and could even make out the curvature of the Earth through their small cabin windows.

The Concorde was powered by four **Rolls-Royce/SNECMA Olympus 593 engines**—two under each wing. The supersonic transport had a range of more than four thousand miles.

During the Cold War, aerospace engineers designed missiles of ever-greater capability. The museum displays a training version of the **Minuteman III intercontinental ballistic missile (ICBM)**, deployed since 1970 by the US Air Force. A three-stage, solid-fuel missile, it has a maximum range of eight thousand miles.

The **guidance system** of the Minuteman III, located in a ring between the upper stage and the nose shroud containing the missile's warheads, is key to its accuracy.

Communications satellites grew in sophistication as they increased in size. This **Sirius FM-4 satellite** represents the first generation of space-based commercial radio service provided by Sirius Radio (now Sirius XM Radio) beginning in 2001. The spacecraft, built as a flight-ready backup, was donated to the museum in 2012. In orbit, the blue solar panel arrays would unfold to span more than seventy-eight feet.

On June 12, 1979, the **Gossamer Albatross**, with long-distance cyclist Bryan Allen as pilot, became the first human-powered aircraft to fly across the English Channel. The modified bicycle that powered the propeller of the seventy-pound aircraft is visible through the transparent Mylar wrapping.

In the 1970s and 1980s, four American space probes—Pioneers 10 and 11 and Voyagers 1 and 2—crossed the asteroid belt to make the first close flyby reconnaissance of the outer planets of the solar system. They returned a wealth of data and images of Jupiter, Saturn, Uranus, and Neptune, along with their moons and rings. Voyager 1, launched in 1977, is still the most distant human artifact from Earth—more than 14.5 billion miles away as of 2022. The museum displays full-scale models of both the Pioneer and Voyager spacecraft.

The museum's full-scale **Pioneer** model was constructed from spare parts assembled in the flight configuration of Pioneer 10. The dish antenna is nine feet wide, and the boom that holds an instrument for measuring magnetic fields is twenty-one feet long.

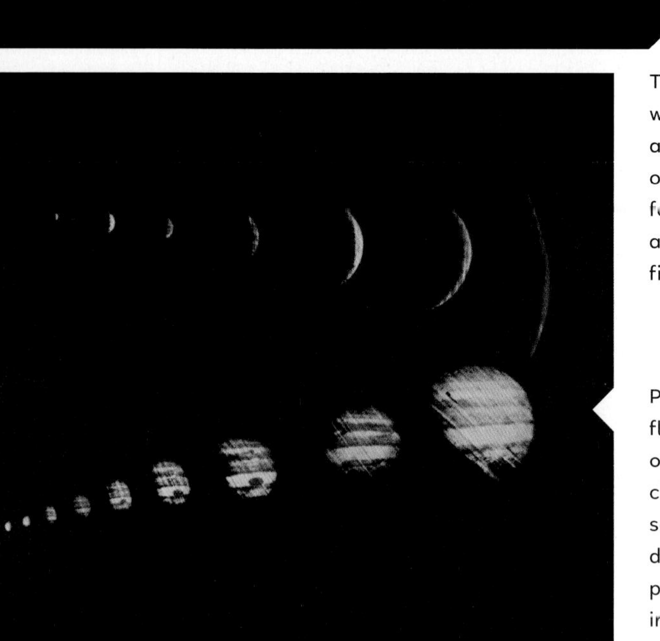

Pioneer 10 was the first spacecraft to fly past **Jupiter**, coming within eighty-one thousand miles of the giant planet's cloud tops on December 4, 1973. This sequence of images was taken on the day of the encounter. Pioneer 11 flew past Jupiter a year later, then went on in 1979 to become the first spacecraft to fly past Saturn.

Both Voyagers carried a "**Sounds of Earth**" record meant to represent life on Earth to any extraterrestrials who might intercept it thousands of years in the future. Selected by a committee chaired by scientist and astronomy popularizer Carl Sagan, the record contained 115 images, spoken greetings in fifty-five languages, natural sounds, and music that ranged from Bach to rock and roll to an Indian raga. This duplicate cover was transferred to the museum in 1978.

Launched in 1977, the twin **Voyagers** both flew past Jupiter and Saturn between 1979 and 1981, while Voyager 2 continued on to encounter Uranus in 1986 and Neptune in 1989. The museum's artifact is a development test model made of facsimile and dummy parts manufactured by the Jet Propulsion Laboratory. Like Pioneer, Voyager had a long boom for holding sensitive instruments away from the main spacecraft body. The shorter boom extending to the upper right holds the spacecraft's nuclear power sources.

Saturn and two of its moons, Tethys (right) and Dione, were photographed by Voyager 1 on November 3, 1980, from a distance of eight million miles. The shadows of Saturn's three bright rings and Tethys are cast onto the planet's cloud tops.

The **Space Shuttle** was the workhorse of the US human spaceflight program for thirty years, from 1981 to 2011. Besides ferrying more people to orbit than any other vehicle before or since, it hauled satellites and cargo and served as a construction base for crews upgrading the Hubble Space Telescope and assembling the International Space Station. Six shuttle vehicles were built, five of which went to space. The museum's orbiter, *Discovery*, flew thirty-nine missions, more than any other. It's displayed at the Steven F. Udvar-Hazy Center alongside the fifty-foot-long telescoping Canadarm manipulator (lower right in the photo) used on *Discovery*'s last six missions to move cargo and astronauts around in space.

Space Shuttle *Discovery*'s **first launch** was on August 30, 1984, from NASA's Kennedy Space Center in Florida.

On three missions in the mid-1980s, the **Manned Maneuvering Unit (MMU)** jet-powered backpack allowed astronauts to "fly" around the shuttle without a safety tether. Bruce McCandless is shown testing MMU serial number 3—now displayed in the Smithsonian—in February 1984.

During the shuttle era, **NASA's astronaut corps** became more diverse, opening to women and minorities. In June 1983, Sally Ride (bottom, second from left), shown here during the STS-41-G shuttle mission in October 1984, became the first American woman in space. On this flight, Kathy Sullivan (to her left) became the first American woman to make a spacewalk.

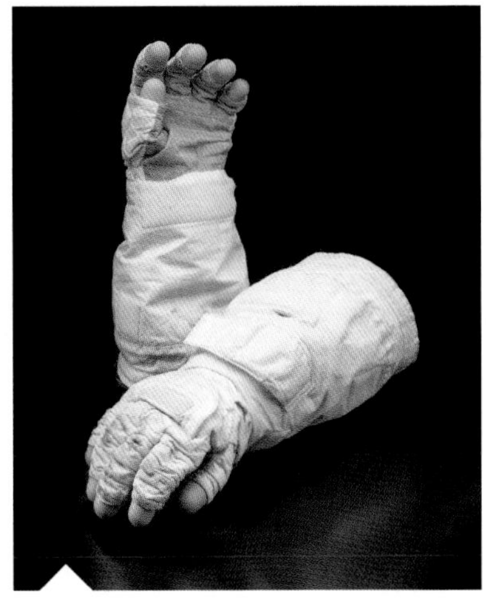

These **gloves** were worn by Kathy Sullivan during her first spacewalk on October 11, 1984. Made of layers of latex, foil, and fabric, the gloves are tough but flexible enough for astronauts to handle tools while working outside their spacecraft.

This full-scale mockup of the **Hubble Space Telescope**—among the most important instruments in the history of science—is on display in the museum's building on the National Mall. Lockheed built this forty-three-foot structure in 1975 to conduct early feasibility studies for the telescope, which was launched aboard the Space Shuttle *Discovery* in April 1990. This test vehicle was used for several purposes, including simulations for developing astronaut maintenance and repair procedures. For more than thirty years, Hubble has given astronomers some of their best views of the most distant objects in the universe.

One of Hubble's most spectacular images this view of a star-forming region in the **Eagle Nebula** was taken by the telescope' wide field and planetary camera 2 (WFPC-2) in 1995. Astronauts returned th camera to Earth after replacing it with ar upgraded version in 2009. WFPC-2 is now in the Smithsonian.

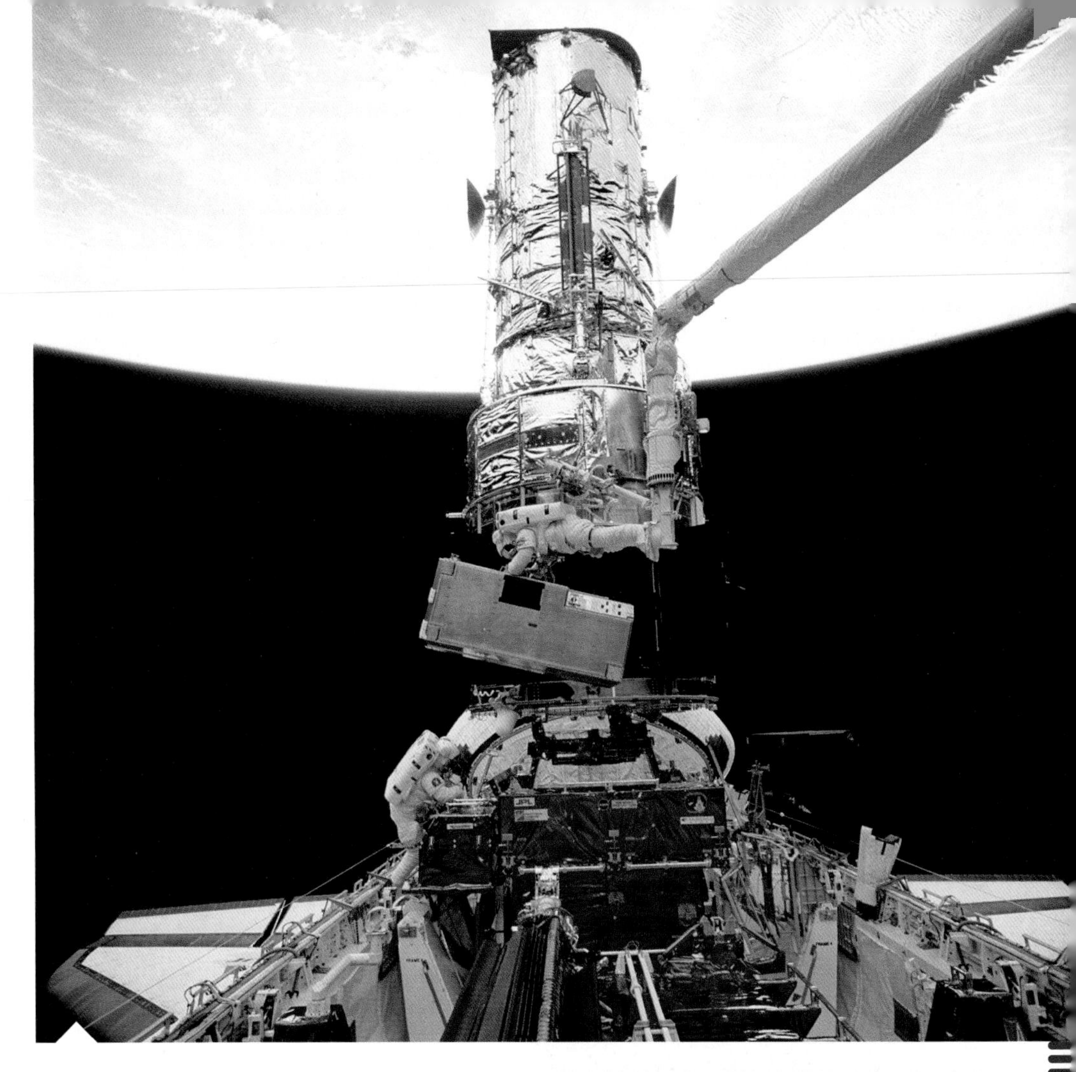

Soon after Hubble was launched in 1990, scientists realized its large primary mirror was slightly misshapen. They designed **COSTAR (Corrective Optics Space Telescope Axial Replacement)**—a long box with nickel-size mirrors that could be placed inside the telescope—to correct for the flaw. Astronauts installed COSTAR and made other fixes during the first Hubble servicing mission in 1993 (above). Then, on the fifth servicing mission in 2009, another astronaut crew replaced it with an instrument that had a built-in correcting mirror. The COSTAR is now in the Smithsonian.

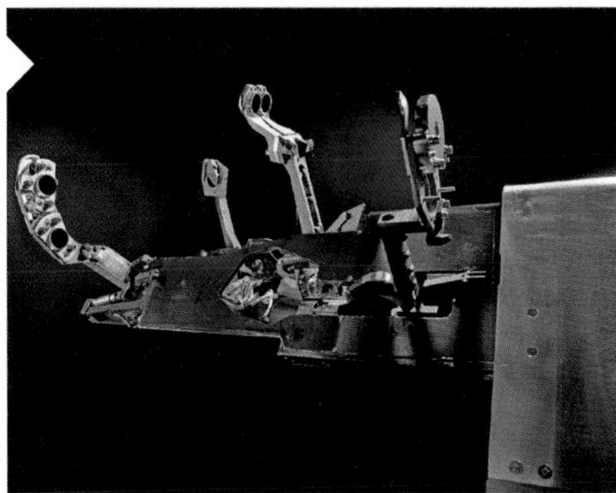

In March 1999, balloonists Bertrand Piccard and Brian Jones completed the first nonstop flight around the world in a free balloon, piloting their **Breitling Orbiter 3** from Switzerland to Egypt in nineteen days. Their 180-foot-tall, aluminum-coated balloon flew as high as thirty thousand feet, where jet stream winds pushed it to speeds of over one hundred miles per hour. Their enclosed gondola is now in the Smithsonian.

In 1999, Cirrus Aircraft introduced an innovative safety feature for small aircraft: a full-plane parachute to ensure a soft landing in case of emergency. Four years later, this **Cirrus SR22, N266CD**, became the first FAA-certified single-engine piston aircraft with a "glass cockpit"—a computerized flight display showing such basic information as altitude and airspeed along with maps and other data. The SR22 became an instant bestseller, energizing the general aviation market. Cirrus later donated N266CD to the museum.

Acclaimed air-show performer **Patty Wagstaff** won the US National Aerobatic Championship three years running, from 1991 to 1993, and was the first woman to win the combined men's and women's title. The custom-built Extra 260 in which she won her first two titles and her flight suit are in the museum.

On July 4, 1997, NASA's ***Mars Pathfinder*** touched down on the Red Planet to kick off a new era in Mars exploration. On board was a one-foot-high, twenty-five-pound prototype mini-rover called *Sojourner* that roamed the rocky Martian surface near the lander for three months. *Sojourner* paved the way for the larger *Spirit* and *Opportunity* rovers that arrived in 2004. Since then, the United States has maintained a continuous robotic presence on Mars: the larger *Curiosity* and *Perseverance* rovers are still exploring the surface, returning key findings about Martian geology.

Instead of retro-rockets, *Mars Pathfinder* used tough, inflatable **airbags** to ensure a soft landing on the Martian surface. An engineering prototype of the airbags is in the Smithsonian along with *Marie Curie*, the flight spare for the *Sojourner* rover (seen here on Mars in 1997, near the deflated airbag poking out from under the lander at bottom right).

The golf-cart-size **Mars Exploration Rover (MER)** Surface System Test-Bed (SSTB) in the museum is nearly identical to the twin rovers, *Spirit* and *Opportunity*, that landed on Mars in January 2004.

The **Lunar Reconnaissance Orbiter (LRO)** has been mapping the Moon at resolutions of less than a meter since 2009. This is a flight spare of LRO's CCD-equipped narrow angle camera (NAC), which is attached to a powerful telescope with a 700mm focal length.

This **LRO view of Earth**, taken from an orbital altitude of about eighty-three miles above the Moon, combines high-resolution black-and-white imagery from the narrow angle camera with lower-resolution color imagery taken by the wide angle camera.

Construction of the International Space Station began in 1998, and the orbital research outpost has been continuously occupied since 2000. Every astronaut's favorite place onboard is the **Cupola,** a multi-windowed observation post for Earth photography and hanging out. Smithsonian visitors will be able to peer inside a full-scale reconstruction of the Cupola mockup that NASA astronauts (like Karen Nyberg, shown above during ISS Expedition 37 in 2013) use for training.

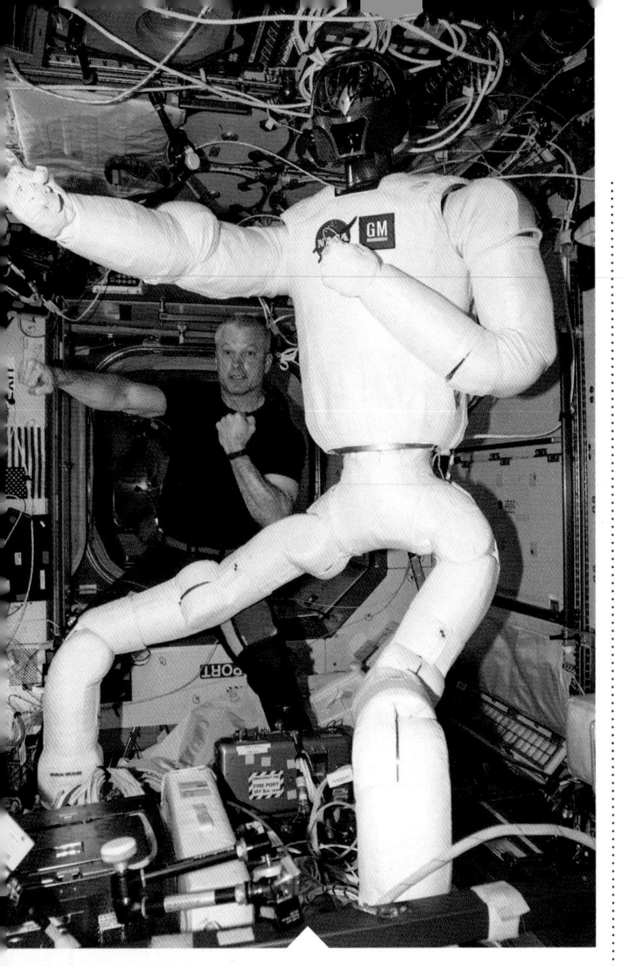

For more than a decade, three **SPHERES** (Synchronized Position Hold Engage and Reorient Experimental Satellites) were onboard the ISS, where the free-floating mini-satellites were used for tests in fields ranging from fluid physics to autonomous robot swarms. All three SPHERES are now back on Earth and in the Smithsonian.

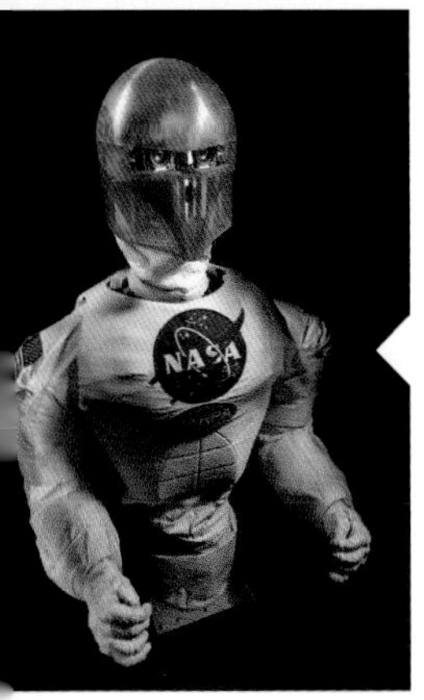

Robonaut was a remotely operated, experimental astronaut "assistant" developed by NASA and the Defense Advanced Research Projects Agency (DARPA) to handle tools and other small objects on the space station. NASA sent two versions to the station. In 2014, Robonaut 2 (above, with astronaut Steven Swanson) was furnished with a pair of legs. Both robots were returned to Earth and are now on display in the Smithsonian.

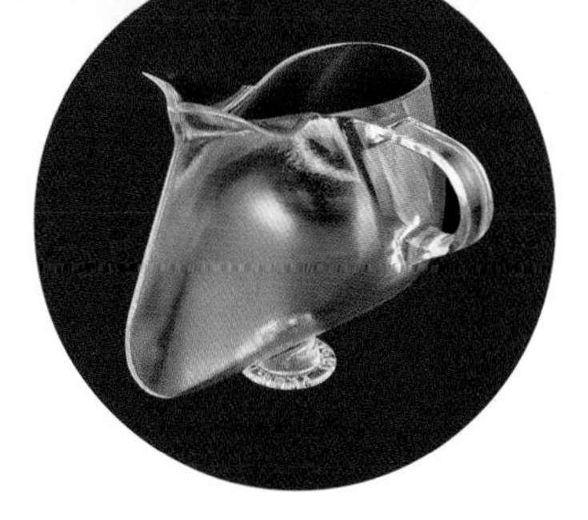

Astronaut Donald Pettit, a veteran of two tours on the International Space Station, worked with IRPI, an Oregon-based company that develops fluid systems for space, to come up with this unique design for a **coffee cup** that doesn't spill in microgravity. Pettit donated one of the cups—a version of which is also available commercially—to the Smithsonian.

INTO THE FUTURE

Science-fiction visionaries from Jules Verne to Steven Spielberg have inspired real aerospace engineers to invent technologies once considered beyond our reach. In the twenty-first century, commercial companies like SpaceX and Rocket Lab are pioneering new approaches to space exploration, while airplane designers aim for new efficiencies using advanced materials and green fuels. At the same time, drones of all shapes and sizes have proliferated, with unpiloted aircraft now capable of performance that the Wright brothers—or even pilots a generation ago—might never have imagined.

(Artwork: *Star Trek Space Scene IV*, Robert T. McCall, ca. 1979)

Made from model train parts and pieces from other model kits, this model of the **alien mother ship** was used in filming *Close Encounters of the Third Kind*, Steven Spielberg's 1977 sci-fi classic. Spielberg conceived the ship's look himself. The sixty-three-inch-long model has tiny, hidden objects added as inside jokes by the model makers. These include a Volkswagen bus, the R2-D2 android, a US mailbox, and a small cemetery plot.

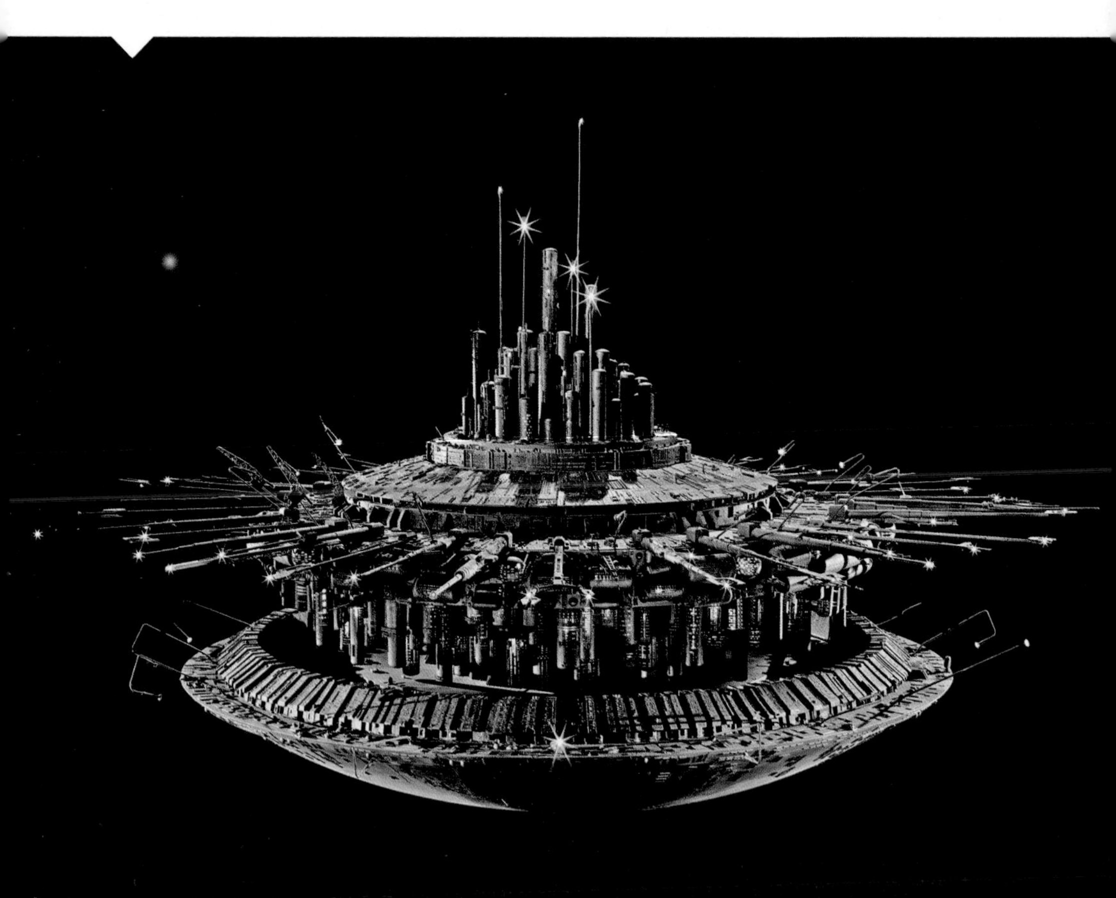

The Smithsonian collection includes this eleven-foot model of starship ***Enterprise*** used in filming the original *Star Trek* series in the late 1960s—one of the most influential shows in television history. The model is constructed mainly of poplar wood, vacuum-formed plastic, rolled sheet metal tubes for the engine pods, and plastic for the main sensor dish and detailing. Paramount Studios donated it to the Smithsonian in 1974.

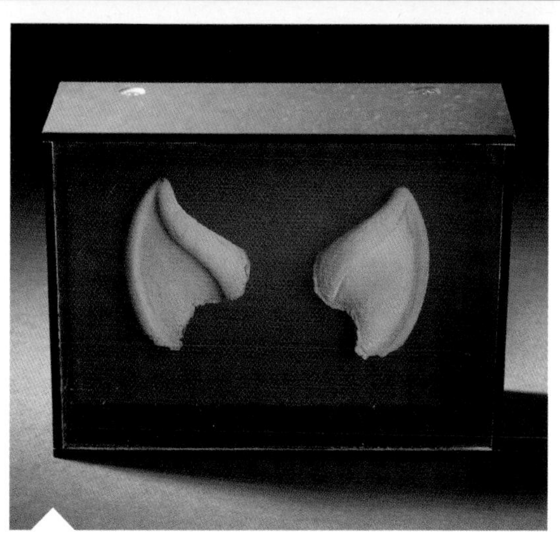

These **latex ears** were worn by actor Leonard Nimoy, who played the half-Vulcan science officer Spock in the original *Star Trek* series. Nimoy made the display case, which he kept in his home until his death in 2015.

This four-inch studio prop from the original *Star Trek* series was used in filming the episode "The Trouble with Tribbles." The synthetic "fur" of this alien tribble is attached to white fabric.

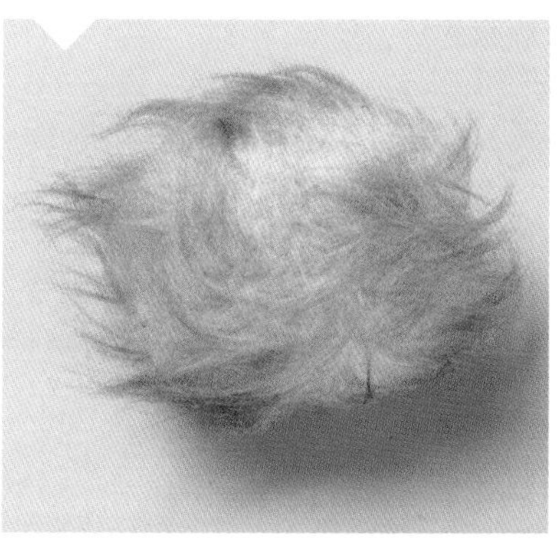

The first privately developed space vehicle to make a suborbital flight, **SpaceShipOne** heralded a new age in which ordinary citizens can fly in space. The rocket-powered spacecraft, built by Scaled Composites and air-launched from a mother ship named *White Knight*, had pivoting wings that "feather" for stable reentry. Pilot Mike Melvill took this ship to an altitude of one hundred kilometers for the first time on June 21, 2004. Two subsequent hundred-kilometer flights earned it the $10 million Ansari X Prize as well as the National Air and Space Museum Trophy for Current Achievement in 2004.

Rocket Lab, an American company founded in 2006, has quickly emerged as a leader in the small-satellite launch market. All the major components of the **Rutherford engine**, developed for the company's Electron rocket, are produced by 3D printing, with an entire engine taking just twenty-four hours to print.

With its Falcon 9 reusable rockets, SpaceX has established a new paradigm for low-cost commercial launch services. Nine **Merlin 1D** engines using liquid oxygen and highly refined kerosene power the rocket's first stage, with an additional Merlin optimized for vacuum conditions on the second stage. The engine donated to the museum was used on three Falcon 9 flights in 2018 and 2019. As the center engine in the first-stage cluster, it was also fired during the booster's dramatic return to Earth (or in this case a drone ship following a 2016 launch), which allows it to be used again.

Falcon 9 rockets also feature black, waffle iron–like "**grid fins**," which deploy from the first stage during descent to provide the aerodynamic control that allows pinpoint landings back at the launch site or on a drone ship. SpaceX donated a grid fin, which was used on a mission to launch the Koreasat-5A satellite in October 2017.

Flown by the US Air Force, Navy, and Marines as well as by the armed forces of other NATO members, the multi-role fifth-generation **F-35 fighter** is a mainstay of modern military air power. The aircraft in the Smithsonian is the first X-35 test vehicle ever built. Originally designated the X-35A, it was modified as the X-35B to include a lift-fan engine for testing short takeoff and vertical landing. The lift-fan propulsion system is displayed next to the aircraft at the Steven F. Udvar-Hazy Center.

Video games have been a popular way of imagining humanity's future in space, including the future of space warfare. Among the most successful video franchises is the *Halo* series created for Xbox beginning in 2001. This **helmet** is part of a costume worn by Master Chief, the main hero of *Halo*, at Xbox public events.

This US Air Force **Predator drone** was one of the first three unmanned aerial vehicles (UAV) to fly operational missions over Afghanistan after the 9/11 terrorist attacks. It went on to fly 196 combat missions in Afghanistan. It was also the first UAV to fire Hellfire missiles in combat.

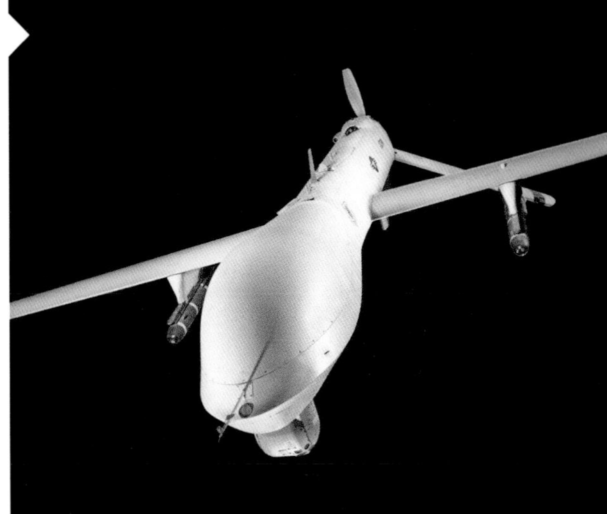

With eight propellers run by solar power, the remotely piloted **Pathfinder Plus** aircraft was used in the 1990s to test high-altitude, long-duration reconnaissance. With a wingspan of 121 feet, it flew as high as eighty thousand feet. The photo is of a test flight over Hawaii in 1998.

In 2012, the remote-controlled **Phantom 1** drone became an instant success among hobbyists. Even a novice drone pilot could operate it, and Chinese manufacturer DJI has gone on to produce more than one million Phantoms. The one on display in the Smithsonian is a gift of the Center for the Study of the Drone at Bard College.

Leonard Niemi's **Sisu** is one of the most successful American competition sailplanes ever flown. In 1964, the engine-less aircraft shattered a symbolic and psychological barrier that had daunted sailplane pilots for years: it flew more than one thousand kilometers on a single flight.

This flight-ready **Sorato rover** was built in 2018 by Japanese aerospace company ispace's Team HAKUTO for the Google Lunar X Prize competition. Although this rover never reached the Moon (the competition was later canceled), ispace is one of several companies now developing similar small rovers for lunar exploration missions in the 2020s.

In 2013, India successfully launched its Mangalyaan orbiter, becoming the first Asian country and the fourth country in the world to reach Mars. **Nandini Harinath**, project manager for the mission at the Indian Space Research Organization (ISRO), donated this sari, which she wore during the spacecraft's departure from Earth orbit on November 30, 2013.

This **view of Mars** taken by India's Mangalyaan spacecraft in September 2014 shows active dust storms in the Red Planet's northern hemisphere.

IMAGE CREDITS

Flickr/Ben Cooper: 111*t*; **Nandini Harinath:** 117*t*; **Imagn:** 57*t*: © *The Columbus Dispatch*—USA TODAY NETWORK; **Indian Space Research Organisation:** 117*b*; **ispace:** 116*t*; **NASA:** 63*l*, 63*mr*, 66*b*: JPL; 69*t*: NASA Dryden Flight Research Center Photo Collection, EC97-44295-51, photo by Carla Thomas; 74*t*: JPL/Caltech; 75*bm*,78*l*&*tr*, 79*br*, 81*tl*, 82*b*: JPL; 84*b*, 94*b*, 95*b*: JPL, 96*t*, 97*tl*, 97*bl*, 98*r*: Hubble Heritage Team/ESA; 99*t*: JSC; 102: JPL; 103*br*: Goddard/Arizona State University; 104*t*; 105*tl*; 105*tr*; 114*t*; **National Archives:** 48*t*; **Jean-Marie Périer/Photo12:** 65*t*; **Rocket Lab:** 110*b*; **Smithsonian Institution Archives:** vi: image #73-7185; vii: Photo by Richard B. Farrar (SI 73-4622); **Smithsonian National Air and Space Museum:** i*t*: Photo by Eric Long (NASM2018-10277), donated by Desert Air Parts Inc.; i*b*: Photo by Dane A. Penland (NASM2013-02525); ii: Photo by Jim Preston (NASM2022-05873); iii*t*: Photo by Eric Long(IMG_4906); iii*b*: Photo by Eric Long (A19280021000); v: Photo by Mark Avino (NASM2020-00501); viii: Photo by Jim Preston (NASM2018-03092); ix: Photo by Jim Preston (NASM2022-05291); 1: *Balloon over City*, (NASM2022-04029), gift of Bella Landauer; 2*t*: Photo by Eric Long (NASM2013-00796); 2*b*: Photo by Dane A. Penland (NASM2016-04281), gift of the Norfolk Charitable Trust; 3*t*: NASM2001-5358; 3*b*: Photo by Eric Long (94-2318); 4*tl*: Photo by Mark Avino (**A19080001000**); 4*tr*: NASM94-5784; 4–5*b*: Photo by Jim Preston (NASM2018-027); 5*t*: Photo by Eric Long (NASM2018-10826), transferred from the Smithsonian Institution to the United States National Museum; 5*b*: NASM73-9000; 6–7: NASM2003-3463; 7*t*: Photo by Jim Preston (NASM2022-00100); 7*b*: Photo by Eric Long (NASM2018-00332); 8*t*: Photo by Eric Long (NASM2005-22899), gift of Curtiss Aeroplane and Motor Company; 8*b*: NASM9A03620; 9*t*: NASM85-14221; 9*m*: Photo by Benjamin G. Sullivan (NASM-26C88A7123882_01); 9*b*: Photo by Dane A. Penland (NASM2014-04336); 10*l*: NASM87-13559; 10–11: Photo by Jim Preston, (NASM2022-04540A), transferred from the US Department of War; 11*bl*: NASM9A00719; 11*br*: NASM9A01205; 12–13: *Ailes Glorieuses*, (A19350067047), gift of Fay Leone Faurote; 14: NASM78-7888; 15*tl*: A19600008000, gift of Mr. James H. R. Cromwell; 15*tr*: Photo by Douglas O. Erickson (A19310061000); 15*b*: Photo by Jason A. Smith (A19620114001); 16: Photo by Eric Long (T8A0161-21); 17*tl*: NASM81-16894; 17*tr*: NASM.1989.0104-m0000389-03520; 17*b*: NASM9A11573-029; 18*t*: Photo by Dane A. Penland (NASM2004-11948), gift of James H. "Cole" Palen; 18*b*: Photo by Eric Long (NASM98-15200); 19*t*: NASM84-6288; 19*b*: Photo by Eric Long (NASM2018-10457), donated by Captain Edward V. Rickenbacker; 20*t*: Photo by Eric Long (NASM2018-10326), transferred from the US Department of War, Bureau of Aircraft Production; 20*bl*: NASM2002-1620; 20*br*: NASM9A00225; 21: Photo by Eric Long (NASM2017-02780), gift of the Arango family; 22–23: *Spirit of St. Louis* by Robert C. Korta, courtesy of *Popular Mechanics*, "The 12 Most Significant Aircraft of All Time," June 1960; 24*t*: NASM87-6058; 24*b*: Photo by David Selsley (A19550027000); 25*t*: Photo by Eric Long (NASM2020-07130); 25*b*: Photo by Eric Long (NASM2011-00582); 26–27: NASM2016-02742, donated by Charles A. Lindbergh; 26*bl*: NASM9A06469; 26*br*: Photo by Mark Avino (A19790148000); 27*bl*: Photo by Carl J. Bobrow (T20130028182); 27*br*: Photo by Anthony S. Wallace (A20030079028); 28*t*: Photo by Eric Long (NASM2021-03401); 28*bl*: NASM92-13721; 28*br*: Photo by Eric Long (NASM2007-13837); 29*t*: Photo by Eric Long (NASM2018-10360), transferred from the US Department of War; 30*t*: NASM7A00449; 30*l*: NASM00160386; 30*r*: Photo by Dane A. Penland (NASM2016-00610), gift of the Daniel and Florence Guggenheim Foundation; 31*l*: Photo by Eric Long (NASM2011-00582); 31*r*: Photo by Dane A. Penland (NASM2005-35502); 32*t*: Photo by Eric Long (NASM2018-10277), donated by Desert Air Parts Inc.; 32*b*: Photo by Jim Preston (NASM2019-00802); 33*tl*: NASM86-533; 33*tr*: Photo by Eric Long (NASM98-15012); 33*b*: Photo by Eric Long (NASM2022-04528); 34: Photo by Eric Long (NASM2018-10363), gift of the Franklin Institute; 35*tl*: NASM79-6354; 35*tr*: A19610155000cp21; 35*br*: Photo by Eric Long (NASM2011-00632); 35*bl*: National Air and Space Museum (SIA-2005-52-02-000001), gift of Amy Morrissey Kleppner; 36*t*: Photo by Eric Long (NASM2018-10064), gift of Eastern Air Lines Inc.; 36*b*: NASM88-9854; 37*t*: Photo by Eric Long and Mark Avino (NASM2006-20909); 37*b*: Photo by Eric Long (NASM2007-13178); 38–39: *Study for Fortresses Under Fire* by Keith Ferris (NASM9A05646); 40*t*: Photo by Eric Long (NASM2018-10346), transferred from the US Air Force; 40*b*: NASM97-17480; 41*tl*: Photo by Eric Long (NASM98-15197); 41*tr*: Photo by Eric Long (NASM2014-02560); 41*bl*: A19700066000; 41*br*: Photo by Craig E. Brunetti (A20160104000); 42*t*: Photo by Eric Long (NASM2014-02561); 42*b*: Photo by Dane A. Penland (NASM2017-01277); 43*b*: NASM9A12266; 43*t*: NASM9A14039-070; 44: Photo by Eric Long (NASM2006-2099); 45*t*: NASM90-69; 45*b*: Photo by Eric Long (NASM2018-10736), transferred from the US Air Force; 46*t*: Photo by Hannah Tippie (A19772838009); 46*r*: Yekaterina Chujkova by Anne Noggle; 46*b*: Photo by Eric Long (NASM2022-04536); 47*t*: Photo by Eric Long (2011-2504), transferred from the US Congress; 47*b*: NASMCW-KT-95; 48*b*: NASM2004-58021-A; 49*t*: Photo by Eric Long and Mark Avino (NASM9A01577); 49*b*: Photo by Mark Avino and Eric Long (NASM2009-12464); 50–51: *Bell X-1 Rocket Powered Aircraft* by Robert C. Korta, courtesy of *Popular Mechanics*, "The 12 Most Significant Aircraft of All Time," June 1960; 52–53: Photo by Eric Long (NASM2015-07319); 52*b*: NASM-USAF-34605AC; 53*t*: Photo by Eric Long (NASM2015-07230); 53*b*: Photo by Eric Long (NASM2015-07223); 54*t*: Photo by Dane A. Penland (NASM2004-25989); 54–55*b*: Photo by Eric Long (NASM2019-01000); 55*t*: Photo by Mark Avino (NASM2019-03322); 55*b*: NASM89-4491; 56*t*: NASM9A02666; 56–57: Photo by Dane A. Penland (A19850806000), gift of Betty Skelton; 57*b*: NASM9A11704; 58*t*: Photo by Eric Long (NASM2006-25353); 58*b*: Photo by Eric Long (A19850563000); 59*t*: Photo by Eric Long (NASM2007-29725); 59*bl*: NASM Archives/GE; 59*br*: Photo by Carl J. Bobrow (A19980293000); 60: Photo by Eric Long (NASM2008-5559); 61*l*: NASM9A08102; 61*t*: Photo by Eric Long (NASM2022-00516); 61*b*: NASM5A35809; 62: Photo by Eric Long (NASM2016-02693), transferred from NASA; 63*tr*: Photo by Eric Long (NASM2018-00323), transferred from NASA; 63*br*: A19670206000, transferred from NASA; 64*t*: Photo by Dane A. Penland (A19730272000), gift of the Boeing Company; 64*m*: Photo by Carolyn Russo (NASM2006-22805), gift of the Boeing Company; 64*b*: NASM Archives/Pan American World Airways; 65*b*: Photo by Dane A. Penland (NASM2004-11706); 66*t*: Photo by Eric Long (NASM2015-7460); 67*l*: Photo by Eric Long (NASM2013-02822); 67*r*: Photo by Eric Long (NASM2006-3298-01); 68–69: Photo by Dane A. Penland (NASM2013-00888); 69*b*: Photo by Doug Dammann (A20020367000); 70: NASM2005-35514; 71*t*: Photo by Dane A. Penland (NASM2013-00885); 71*m*: Photo by Dane A. Penland (NASM9A08353); 71*b*: A20060185000; 72–73: *Study for Lunar Landscape Mural* by Chesley Bonestell (A20200171000); 74*b*: Photo by Eric Long (NASM2018-00854), transferred from NASA; 75*t*: NASM9A14513; 75*tm*: Photo by Eric Long (NASM2006-19886); 75*b*: Photo by Carl J. Bobrow (A19700114000); 76*t*: A19740129000, gift of Norman Rockwell Family Agency; 76*b*: Photo by Mark Avino (NASM2005-29580); 77*tl*: NASM-NASA-65-HC-359; 77*tr*: Photo by Dane A. Penland (NASM2014-05409); 77*b*: Photo by Mark Avino (NASM2019-10008), transferred from NASA; 78*br*: Photo by Eric Long (A19980009000); 79*tl*: Photo by Eric Long (NASM2005-6758); 79*tr*: Photo by Mark Avino (A19700102000); 79*bl*: Photo by Jim Preston (NASM2018-02095), transferred from NASA; 80: Photo by Eric Long (NASM2016-03148), transferred from NASA; 81*tr*: Photo by Mark Avino (NASM2020-00046); 81*bl*: Photo by Eric Long (A19830142000); 81*br*: Photo by Jim Preston (NASM2019-05863); 82*t*: Photo by Mark Avino (A19790215000), transferred from NASA; 83*tl*: Photo by Mark Avino (NASM2016-02044); 83*tr*: National Air and Space Museum and Smithsonian Institution Archives; 84*t*: NASM9A01511; 85*tl*: Photo by Eric Long (NASM2005-22900); 85*tm*: Photo by Eric Long (NASM2004-51132.02); 85*tr*: Photo by Mark Avino (A19772551000); 85*br*: Photo by Eric Long (NASM2006-18605); 86–87: *Dale A. Gardner, Space Shuttle Mission 51-A* by George D. Guzzi Jr. (NASM2014-06837), gift of George D. Guzzi Jr.; 88*t*: NASM2000-9716; 88*b*: Photo by Eric Long (NASM2020-11443); 89*t*: NASM-9A19294, courtesy of Boeing Company; 89*b*: NASM-9A11647, courtesy of United Airlines; 90: Photo by Dane A. Penland (NASM2019-05285); 91*l*: Photo by Eric Long (NASM2005-4060); 91*tr*: NASM-7A09499, photo by Arthur Ernest Gibson, courtesy of Jacqueline Webby; 91*br*: Photo by Dane A. Penland (NASM2019-05279); 92*l*: Photo by Eric Long (NASM2011-01548); 92*r*: Photo by Eric Long (NASM2006-1767); 93*t*: Photo by Eric Long (NASM2021-000[illegible]); 93*b*: Photo by Dane A. Penland (NASM2013-02069); 94*t*: Photo by Eric Long (NASM2016-00084); 95*t*: NASM9A03167; 95*m*: Photo by Eric Long (NASM2005-22901); 96*b*: Photo by Dane A Penland (NASM2013-02525); 97*tr*: Photo by Dane A. Penland (SI-2006-104), transferred from NASA; 97*br*: Photo by Dane A. Penland (A19960009000), transferred from NASA Johnson Space Center; 98*l*: Photo by Eric Long (NASM2013-02349); 99*b*: Photo by Eric Long (NASM2013-02349); 100*t*: Courtesy of Breitling; 100*m*: Photo by Dane A. Penland (A19990257000), gift of Breitling; 100*b*: A19990257000, courtesy of Cirrus Aircraft and Patrick Sniffen; 101*t*: Photo by Carolyn Russo (NASM94-9327); 101*b*: Photo by David Selsley (A19930403000); 103*t*: Photo by Mark Avino (NASM2020-00501); 103*bl*: Photo by Eric Long (NASM2016-00444); 104*b*: Photo by Jim Preston (NASM2022-05063); 105*bl*: Photo by Eric Long (NASM2020-00384); 106–107: *Star Trek Space Scene IV* by Robert T. McCall (A19810290000), photo still from *Star Trek: The Motion Picture*, courtesy of CBS Broadcasting Inc.; 108: Photo by Eric Long (A19790906000); 109*t*: Photo by Mark Avino (NASM2005-34607); 109*bl*: Photo by Eric Long (NASM2021-06493), gift of the Nimoy Family, in honor of Beit T'Shuvah and the Leonard Nimoy COPD Research Fund at UCLA; 109*br*: Photo by Benjamin G. Sullivan (A19850351000), gift of Paramount Pictures Corp.; 110*t*: Photo by Eric Long (NASM2015-05848); 112–113*t*: Photo by Dane A. Penland (NASM2013-00893); 112*b*: Photo by Mark Avino (NASM2022-01941); 113*b*: Photo by Eric Long (NASM2017-03151); 114*b*: Photo by Dane A. Penland (A20070057000); 115*t*: Photo by Dane A. Penland (NASM2005-24790), gift of Philip J. Baugh; 115*b*: Photo by Mark Avino (NASM2022-04663); 116*b*: Photo by Emily M. Smithberger (A20220323000); **Spaceware:** 105*br*; **SpaceX:** 111*b*; **Ted Stryk/Planetary Society:** 83*b*.